The Essential Guide to Passing the Water Resources and Environmental Civil PE Exam Written in the form of Questions

160 CBT Questions Every PE Candidate Must Answer

First Edition

Jacob Petro
PhD, PMP, CEng, PE

Request latest Errata, or add yourself to our list for future information about this book by sending an email with the book title, or its ISBN, in the subject line to:
Errata@PEessentialguides.com
or
Info@PEessentialguides.com

The Essential Guide to Passing the Water Resources and Environmental Civil PE Exam Written in the form of Questions

160 CBT Questions Every PE Candidate Must Answer

First Edition

Jacob Petro

PhD, PMP, CEng, PE

PE Essential Guides

Hillsboro Beach, Florida

> **Report Errors For this Book**
>
> We are grateful to every reader who notifies us of possible errors. Your feedback allows us to improve the quality and accuracy of our products.
>
> Report errata by sending an email to Errata@PEessentialguides.com

The Essential Guide to Passing the Water Resources and Environmental Civil PE Exam Written in the form of Questions – 160 CBT Questions Every PE Candidate Must Answer

First Edition - Print 1.4.2

© 2024 Petro Publications LLC. All rights reserved.

All content is copyrighted by Petro Publications LLC and its owners. No part, either text or image, may be used for any purpose other than personal use. Reproduction, modification, storage in retrieval system or retransmission, in any form or by any means, electronic, mechanical, or otherwise, for reasons other than personal use, without prior permission from the publisher is strictly prohibited.

For written permissions contact: Permissions@PEessentialguides.com

For general inquiries contact: Info@PEessentialguides.com

Imprint name: PE Essential Guides

Company owning this imprint: Petro Publications LLC. Established in Florida, 2023.

ISBN: 979-8-9891857-6-4

Release History

Date	Edition No.	Description and Update
March 2024	1	

Disclaimer

The information provided in this book is intended solely for educational and illustrative purposes. It is important to note that the technical information, examples, and illustrations presented in this book should not be directly copied or replicated in real engineering reports or any official documentation.

While there may be resemblances between the examples in this book and real structures, users must exercise caution and conduct comprehensive verification of all information before implementing it in any practical setting. The author and all affiliated parties explicitly disclaim any responsibility or liability arising from the misuse, misinterpretation, or misapplication of the information contained in this book.

Furthermore, it is essential to understand that this book does not constitute legal advice, nor can it be considered as evidence or exhibit in any court of law. It is not intended to replace professional judgment, and readers are encouraged to consult qualified experts or seek legal counsel for any specific legal or technical matters.

By accessing and utilizing the information in this book, readers acknowledge that they do so at their own risk and agree to hold the author and all affiliated parties harmless from any claims, damages, or losses resulting from the use or reliance upon the information provided herein.

Important Information About Printing and Quality Assurance

Most of our publications are printed through third-party services. While we have full trust in our partners, we acknowledge that occasional minor issues, such as missing pages or other errors, may arise. If you encounter any such issues with your copy, please do not hesitate to contact us via any of the emails listed on the copyright page, and we will be happy to assist you in resolving the matter.

Preface

General Information about the Book

This book is designed to help civil engineers pass the NCEES exam with its 2024 updated specifications, which is a prerequisite for obtaining the professional engineering PE license in the United States 2024 onwards. This book is tailored to provide you with comprehensive knowledge, detailed examples, and step-by-step solutions with ample graphics that are directly related to the subjects covered by the NCEES exam.

In this book, you will find an extensive collection of civil engineering problems that are carefully selected to build your knowledge, skills, and ability to apply fundamental principles and advanced concepts in the field of civil engineering. These problems are accompanied by detailed explanations, diagrams, and equations to help you understand the underlying principles and solve the problems efficiently and accurately.

Whether you are a recent graduate, an experienced engineer, or a professional who wants to obtain the engineering license in the United States, this book will prepare you for the exam and equip you with the necessary tools to succeed.

The book is structured in a way such that it provides the reader with a comprehensive understanding of the core topics that are covered by the NCEES exam 2024 specifications – which have more in-depth focus compared to the previous versions of the same exam.

The book provides the reader with a full coverage and understanding for the NCEES relevant exam topics and possible question scenarios during a real test situation. If certain topics or methods were not covered in this book, the book method of presentation will ultimately guide you on how to find the solution you seek on your own, know where to find it, and provide the solution on a timely manner that saves you time during the real exam.

The questions in this book are neither easy nor difficult. They are constructive and creative in nature. They have been authored in a way to have you remember the core concept of the engineering topic you seek. They are designed to challenge engineers to think critically and apply their knowledge to exam and real-world scenarios. These questions require a deeper level of analysis and understanding than simple recall of information. They may involve multiple steps or require the engineer to consider different perspectives or solutions, while they can be difficult and require significant amount of effort and research to solve them.

Reasons I wrote this Book

I decided to author this book because I have a strong passion for engineering. I have a deep interest and understanding of civil engineering, and I wanted to share this knowledge and passion with others.

I am also very enthusiastic about engineering, and this has allowed me to explore various concepts and develop unique perspectives.

By writing this book I hope I can inspire others to pursue and improve on their career in engineering, help them pass the NCEES exam, improve on their skills and advance their knowledge in this field and provide them with the tools they need to succeed.

Lastly, the energy and enthusiasm I have, and I brought into this work is infectious and I wanted to channel this energy into this fascinating project and share it with others. I strongly believe this book will be a valuable resource for anyone interested in learning more about civil engineering.

Acknowledgement and Dedication

I would like to thank the readers of this book, who I hope will find it informative, engaging, and thought-provoking. It is my sincere hope that this book will inspire others to pursue their own passion, and that it will serve as a valuable resource for all those interested in the field of engineering.

Table of Contents

Preface vii

    General Information about the Book vii
    Reasons I wrote this Book vii
    Acknowledgement and Dedication viii

Introduction 1

    General description of this Book 1
    Book Structure 1

Book Parts 2

About the Exam 3

    General information 3
    Dissecting the Exam 3

How to use this Book 4

Which References to Own 5

2024 Exam Specifications 5

Map of Problems Presented 9

    Project Planning 9
    Soil Mechanics & Materials 9
    Hydraulics: Closed Conduit & Open Channel 9
    Hydrology, Surface Water, Groundwater & Wells 10
    Analysis & Design: Water, Wastewater & Water Quality 11

Project Site Work 12

Problems & Solutions 13

Project Planning 15
Soil Mechanics & Materials 31
Hydraulics: Closed Conduit & Open Channel 49
Hydrology, Surface Water, Groundwater & Wells 81
Analysis & Design: Water, Wastewater & Water Quality 99
Project Site Work 145

References 167

Bibliography 167

Permissions 169

Index 170

About the Author

Dr. Petro is a professional engineer and a business leader with over 20 years of experience in leading and growing engineering companies. Throughout his career, he worked with some of the most prestigious engineering firms. With a vast background in design and construction, he earned a reputation for delivering innovative and cutting-edge projects throughout his career.

Dr. Petro is a civil engineer, he holds a Doctorate degree, he has earned a Professional Engineering (PE) license as well as Chartered Engineer (CEng) certification. Additionally, Dr. Petro has earned a Project Management Professional (PMP) certification, further demonstrating his expertise in managing complex projects. Over the years, he successfully led and managed teams of engineers, designers, and other professionals, overseeing complex projects from conception to completion.

Throughout his career, Dr. Petro designed and delivered numerous innovative and interesting projects that have contributed significantly to various industries he has worked in. His passion for engineering and business has driven him to publish several papers and articles in industry-leading journals and magazines. His work has been recognized as state-of-the-art and has been referenced by many industry professionals.

As an international civil engineer who worked across the globe, Dr. Petro brings an interesting perspective to the table. He has a deep understanding of how civil facilities and structures work and how to optimize them for maximum efficiency and safety. His ability to communicate complex engineering concepts to both technical and non-technical stakeholders has been key to his success.

This page is intentionally left blank

Introduction

General description of this Book

The Essential Guide to Passing The Water Resources and Environmental Civil PE Exam is a guide designed in the form of questions. It aims to achieve a comprehensive and good coverage for the water, wastewater, and environmental engineering. It attempts to cover numerous exam scenarios with the use of questions. It also explains theory with examples when needed – Look for the book sign 📖 for questions that provide theory explanation.

Also, given the diverse nature of the water resources and environment field, it is unlikely that all relevant information can be found in a single source. We have done this for you (check the references section). We carried out the necessary amount of research and consequently, we have gathered information from multiple sources to provide you with a good coverage for the water resources and environmental engineering NCEES required knowledge areas.

Although the guide is designed with the use of 160 questions, some questions require two outcomes for a solution making them worth two questions not one. Those questions were intentionally designed this way so that a certain concept can be delivered which is intertwined with another, also, to ensure that the strategy of comprehensive coverage, which is the backbone strategy behind this book, is achieved. All in all, our aim behind authoring this book is to make it a one stop shop for passing the Water Resources and Environmental Civil PE exam.

The answers to the questions are detailed and they are well referenced. Some answers provide several methodologies in their presentation for the readers' benefit. Questions and answers are accompanied by graphics and detailed step-by-step explanations to help engineers understand concepts better. Not only this, the aim of providing such detailed answers is to help engineers understand all methods and apply them during the exam so they are better equipped with all possible exam scenarios.

Book Structure

The 2024 water resources and environmental PE exam specifications consist of twelve knowledge areas. Those knowledge areas are presented in the relevant section of this book.

For ease of reference, and in order to ensure a clear, and a complete coverage for all possible question scenarios, those twelve knowledge areas are grouped into six parts/sections in this book along with the number of questions, and they are as follows:

Book Part/Section	No
Project Planning	12
Soil Mechanics & Material	22
Hydraulics: Closed Conduit & Open Channel	32
Hydrology, Surface Water, Ground Water & Wells	24
Analysis & Design: Water, Wastewater & Water Quality	50
Project Site Work	20

The number of questions in each part was carefully determined in a way that speaks to the expected number of questions in a real exam, also, they have been made as such ensuring good coverage for the intended material behind relevant knowledge areas for each part.

Book Parts

As described earlier, the book problems are arranged in six parts as presented in the Problems and Solutions Section. The coverage for each of the six parts is summarized in the following paragraphs:

The *Project Planning Part* covers topics that are mostly relevant to cost estimating and scheduling, economic analysis such as present worth and lifecycle cost, and other topics that are relevant to the project management knowledge area of the 2024 exam specifications.

The *Soil Mechanics and Material Part* contains questions from the relevant knowledge area. Soil sampling, geotechnical evaluation, slope evaluation, consolidation and laboratory tests along with pipes and pipe material evaluation and many other relevant topics are covered in this part.

The *Hydraulics – Closed Conduit and Open Channel Part* covers the principles of hydraulics and consists of numerous questions with good coverage relevant to fluid mechanics, pressure evaluation and pressure on gates, viscosity, principles of fluid flows and flow characteristics, the impulse momentum principle, head losses and various scenarios for head losses in pipes, culverts and channels, pipes flowing full or partial, pump cavitation and NPSH applications, pipe network – parallel, loop and series, energy diagram and critical and normal depth, hydraulic jumps and the Gradually Varied Flow GVF, culverts and culvert flows and head losses and gutter flow, etc.

The *Hydrology, Surface Water, Ground Water & Wells Part* covers the analysis and subsequent design for storm and storm events, runoff analysis methods, rainfall intensity, design of ponds, sheet flow analysis, hydrographs, rainfall gauging stations, depletion and infiltration, soil loss and erosion. It also covers for groundwater and wells, such as aquifers, hydraulic conductivity, Darcy's law, and Dupuit's formula., aquifer layers and aquifer analysis.

The *Analysis & Design – Water, Wastewater & Quality Part* contains interesting problems and theory coverage for various scenarios starting with mass conversions and continuity law, oxygen demand and oxygen saturation and deficit along with deoxygenation using different methods. Aeration and reaeration rates, stream oxygen deficits, and time to critical deficit. The Monod's kinetics for growth modelling as well. Drinking Water Equivalent Levels DWELs, and the Chronic Daily Intake for contaminated waters.

There is also good coverage and theory explanation for the wastewater and collection, waste water testing and the various reactions that take place in plants, head losses due to filtration and hydraulics, primary treatment analysis and design for clarifiers, sedimentation tanks, the various processes for treatment explained with examples and more in depth, loading rates and the mixed liquor suspended solids, sludge and sludge calculations, the Return Activated Sludge RAS calculations, the dissolved air flotation, trickling filters, nitrification and denitrification and phosphorus removal explained. Solid treatment and handling, digestion, aerobic and anaerobic treatment, advanced treatment and the use of methanol and acetate.

When it comes to water distribution systems, there is a good coverage as well for distribution and treatment, treatment processes, storage, mixing design and flocculator design, adsorption methods and calculations, air stripping methods and calculations, hardness and softening methods and calculations, membrane filtration, odor control and chlorination and dechlorination methods and calculations.

The *Project Site Work Part* contains questions on excavations and mass haul diagram theory, site works and layout, soil prevention losses and erosion control and erodibility, safety and construction safety, construction activities and operations activities examples, roadside drainage channels' calculations. It also has questions on basic horizontal and vertical curves, retaining walls and lateral pressures with hydraulics involved, etc.

For ease of reference, knowledge areas covered in each of these parts are presented at each part's page break. Also, for better and quick understanding for topics covered in those parts, you can refer to the Map of Problems Presented section.

Although this book is divided into six distinct parts, it is important to recognize that these parts are not entirely separate from each other. In fact, and due to the nature of the water resources and environmental engineering, these parts share common themes and ideas could overlap.

About the Exam
General information
The NCEES PE exam is a rigorous exam that is administered in two sessions, a morning session, and an afternoon session. The morning session is four hours long that used to focus on broader, foundational concepts in the engineering field along with a wide range of engineering problems compared to the afternoon session. This session is replaced in 2024 in the new exam specifications with more in-depth topics that are relevant to the water resources and environmental exam along with some relevant project planning, project sitework and other relevant breadth questions as explained here.

The afternoon session is also four hours long and is generally more focused on specific areas of expertise.

Both the morning and the afternoon sessions with the 2024 specifications carry a similar weight when it comes to importance and depth of coverage.

Both sessions consist of multiple-choice questions, point-and-click, drag-and-drop, or fill in the blank type. Candidates are typically required to demonstrate their ability to analyze and solve complex engineering problems during those sessions.

All in all, this exam is designed to test not only one's knowledge and technical skills, but also their ability to think critically and work under pressure.

Dissecting the Exam
The exam consists of 80 questions presented in two sessions. Candidates are given a total of 480 minutes to solve those questions. This means that each question is allotted an average time of six minutes. It is very important to keep in mind that some questions during the real exam will take only one minute to solve while others could take up to ten minutes to complete. To prepare for the exam, it is crucial to practice solving questions that have longer duration and that

are more difficult, which is what this book aims to provide.

I would like to provide a reflection from my own experience sitting for the NCEES PE exam: During the exam, I was thoroughly prepared to confidently answer all questions within the allotted time, and even expected to complete the exam at a faster pace. However, I regretfully failed to study two specific chapters of a certain manual that I had assumed would have a lower probability of being included in the exam. To my dismay, two difficult questions emerged from these chapters, leaving me with no alternative but to study the relevant material from the codes and manuals provided eating away from the exam time. Consequently, I spent approximately 20-30 minutes ensuring that my answers were accurate. This unexpected event significantly impacted the remainder of my allotted exam time, and although I ultimately passed, it was an avoidable situation.

This experience is one of the reasons why I strived to provide a comprehensive coverage of all possible topics and scenarios in this book, such that exam candidates do not have to go through this experience.

How to use this Book

The questions presented in this book are designed with a mix of varying lengths. Some may only take a minute or two to answer, while others may require up to 10 or 15 minutes. This design has been done as such intentionally, as it reflects the format of the actual PE exam with more questions that are longer and more difficult than the exam.

By practicing with questions of varying lengths, candidates will be better prepared to manage their time during the exam. They will learn to quickly identify fewer complex questions and move through them efficiently, while also having the skills to tackle the more time-consuming questions effectively.

It is important to note that practicing only short questions, or questions that are six minutes long, may not be enough to fully prepare for the exam.

Furthermore, the variety of questions' lengths presented in this book helps keep one's mind engaged and challenged. It can be easy to become bored or disengaged when faced with a series of similar questions, but by mixing up the lengths, candidates will be forced to stay focused and adapt to the different types of questions.

Therefore, it is important to practice longer and more difficult questions in addition to the shorter ones. This will help develop one's ability to think critically, analyze complex problems, and apply their knowledge to solve them. It will also help build one's endurance and focus, which is critical for success on this exam.

Moreover, it is important to note that it is okay to spend more time on difficult questions or even shorter ones during the practice sessions. This will help identify weaknesses and areas of improvement.

As a final note, it is important to view this book as a textbook as it contains, not only straightforward questions and answers, but comprehensive explanations and guidance and wide coverage to the most complex exam problems with detailed elaborations of both questions and answers. It includes theory explanation as well and a wide range of references that you can use at your own pace.

Which References to Own

The water resources and environmental engineering civil PE exam requires you to thoroughly study three references, those references can be downloaded from the web, and they are listed in the references section of this book.

2024 Exam Specifications

Effective April 2024 the exam will focus more on the depth part with some general, but relevant, items from the planning and project management and geotechnical knowledge areas.

This guide takes this change into account and the questions were authored with the 2024 exam specifications in mind.

The 2024 exam specifications are presented in the following three pages for ease of reference. Also, parts of the specification are presented in the Problems & Solutions relevant page breaks for each part denoting which areas are covered in that part.

SN	Knowledge Area	Expected Number of Questions
1	**Project Planning** A. Quantity take-off methods B. Cost estimating C. Project schedules D. Activity identification and sequencing E. Economic and sustainability analysis (e.g., present worth, lifecycle costs, comparison of alternatives)	4-6
2	**Soil Mechanics** A. Lateral earth pressure B. Soil consolidation and compaction C. Bearing capacity D. Settlement E. Slope stability	3-5
3	**Materials** A. Soil classification and boring log interpretation B. Soil properties (e.g., strength, permeability, compressibility, phase relationships) C. Concrete (e.g., nonreinforced, reinforced) D. Piping materials E. Material test methods and specification conformance	4-6
4	**Analysis and Design** A. Mass balance B. Hydraulic loading C. Solids loading (e.g., sediment loading, sludge) D. Hydraulic flow measurement	6-9
5	**Hydraulics – Closed Conduit** A. Energy and/or continuity equation (e.g., Bernoulli, grade line analyses, momentum equation) B. Pressure conduit (e.g., single pipe, force mains, Hazen-Williams, Darcy-Weisbach, major and minor losses) C. Pump application and analysis, including wet wells, lift stations, and cavitation. D. Pipe network analysis (e.g., series, parallel, loop networks)	7-11
6	**Hydraulics – Open Channel** A. Open-channel flow B. Hydraulic grade lines and energy dissipation (e.g., plunge pool, drop structure, culvert outlet) C. Stormwater collection and drainage (e.g., culvert, stormwater inlets, gutter flow, street flow, storm sewer pipes) D. Sub- and supercritical flow	7-11

7	**Hydrology** A. Storm characteristics (e.g., storm frequency, rainfall measurement, distribution) B. Runoff analysis (e.g., rational and SCS/NRCS methods) C. Hydrograph development and applications, including synthetic hydrographs D. Rainfall intensity, duration, frequency, and probability of exceedance E. Time of concentration F. Rainfall and stream gauging stations G. Depletions (e.g., evaporation, detention, percolation, diversions) H. Stormwater management and treatment (e.g., detention and retention ponds, infiltration, swales, constructed wetlands)	8-12
8	**Groundwater and Wells** A. Aquifers B. Groundwater flow C. Well and drawdown analysis	4-6
9	**Surface Water and Groundwater Quality 5–8** A. Stream degradation and oxygen dynamics B. Total maximum daily load (TMDL) (e.g., nutrient contamination, DO, load allocation) C. Biological and chemical contaminants	5-8
10	**Drinking Water Distribution and Treatment** A. Drinking water distribution systems B. Drinking water treatment processes C. Present, short-term, and long-term demands D. Storage E. Sedimentation F. Coagulation and flocculation G. Membrane processes and media filtration H. Disinfection, including disinfection byproducts I. Hardness and softening J. Other treatment (e.g., ion exchange, carbon adsorption, ozone, UV, specific constituent removal)	6-9
11	**Wastewater Collection and Treatment 7–11** A. Wastewater collection systems (e.g., lift stations, sewer networks, infiltration, inflow, smoke testing, maintenance, odor control) B. Wastewater treatment systems C. Preliminary treatment D. Primary treatment E. Secondary treatment (e.g., physical, chemical, biological processes) F. Nutrient removal G. Solids treatment, handling, and disposal H. Disinfection I. Advanced treatment (e.g., advanced oxidation process, effluent filtration, adsorption, reclaimed water)	7-11

12	**Project Sitework 9–14**	9-14
	A. Excavation and embankment (e.g., grading, cut and fill)	
	B. Construction site layout and control	
	C. Temporary and permanent soil erosion and sediment control (e.g., construction erosion control and permits, sediment transport, channel/outlet protection)	
	D. Impact of construction on adjacent facilities	
	E. Safety (e.g., construction, roadside, work zone)	
	F. Basic horizontal and vertical curve elements	
	G. Retaining walls	
	H. Construction methods	

Map of Problems Presented

Project Planning

Problem 1.1 Project Cost Analysis	**Problem 1.2** Project Activity Sequencing
Problem 1.3 Budgetary Cost Acceptable Range	**Problem 1.4** Benefit Cost Analysis
📖 (✤) **Problem 1.5** Net Present Value Analysis	**Problem 1.6** Resources Histogram
📖 **Problem 1.7** Mass Haul Diagram	**Problem 1.8** Profit and Loss
Problem 1.9 Capitalized Costs	**Problem 1.10** Rate of Concrete Pouring
Problem 1.11 Bond Effective Interest	**Problem 1.12** Depreciated Rate

Soil Mechanics & Materials

Problem 2.1 Pressure Under Footing	**Problem 2.2** Foundation Settlement
📖 **Problem 2.3** Effective Stress Over Time	**Problem 2.4** Optimum Moisture Content
Problem 2.5 Slope Stability	**Problem 2.6** Construction Settlement
Problem 2.7 Bearing Capacity for a Square Foundation	(✤) **Problem 2.8** Soil Properties (1)
Problem 2.9 Dam Site Rock Quality	**Problem 2.10** Soil Properties (2)
Problem 2.11 Soil Shear Strength	**Problem 2.12** Settlement in Clay
Problem 2.13 Boreholes Locations	📖 **Problem 2.14** Soil Moisture Content
Problem 2.15 Soil Classification System (1)	**Problem 2.16** Soil Classification System (2)
Problem 2.17 Concrete Mix Design	**Problem 2.18** Soil Permeability Testing
Problem 2.19 Cement Type	**Problem 2.20** Hydrostatic Pressure Test
Problem 2.21 Pipe Material (1)	**Problem 2.22** Pipe Material (2)

Hydraulics

Problem 3.1 Water Surface Elevation	**Problem 3.2** Head Losses
Problem 3.3 Elevation of Water Surface in Two Reservoirs	**Problem 3.4** Vertical Force at a Joint of a Jet
(✤) **Problem 3.5** Water Discharge External Forces	**Problem 3.6** Gate under Pressure
Problem 3.7 Hazen-Williams Coefficient Determination	**Problem 3.8** Increase in Pipe Flow due to Diameter Increase

Problem 3.9￼Water Hammer Pressure	Problem 3.10￼Head Increase Required by a Pump
Problem 3.11￼Packed Bed Design	Problem 3.12￼Pump Required Power
Problem 3.13￼Pump Upgrade	📖 Problem 3.14￼Pump Elevation
Problem 3.15￼Loop Network Flow Calculation	Problem 3.16￼Parallel Network Flow Calculation
Problem 3.17￼Seawater Canal	Problem 3.18￼Water Channel
Problem 3.19￼Flow in a Gutter	Problem 3.20￼Culvert Inlet Headwater Elevation
Problem 3.21￼Culvert Outlet Headwater Elevation	Problem 3.22￼Hydraulic Jump
Problem 3.23￼Energy Loss	Problem 3.24￼Half Circular Channel
Problem 3.25￼Pipe Flow Depth	Problem 3.26￼Culvert Flow
Problem 3.27￼Critical Slope	Problem 3.28￼Hydraulic Radius Composite Channel
Problem 3.29￼Depth of Flow	📖 Problem 3.30￼Gradually Varied Flow
Problem 3.31￼Roughness Coefficient	Problem 3.32￼Culvert Flow Velocity

Hydrology, Surface Water, Groundwater & Wells

📖 Problem 4.1￼Watershed Rainfall Depth	📖 Problem 4.2￼Precipitation Methods
Problem 4.3￼Retention Pond Sizing	Problem 4.4￼Shallow Flow
Problem 4.5￼Travel Time for Shallow Flow	Problem 4.6￼Retention Basin Design
Problem 4.7￼Composite Curve Number	📖 Problem 4.8￼10-Year Peak Flow
Problem 4.9￼Hydrograph Excess Rain	Problem 4.10￼Hydrograph Development
📖 Problem 4.11￼Hydrograph Shape	Problem 4.12￼Rainfall Intensity
Problem 4.13￼Rain Gauge Precipitation Estimation	Problem 4.14￼Exceedance Probability
Problem 4.15￼Horton Infiltration Capacity Curve	Problem 4.16￼Hydrologic Budget (1)
Problem 4.17￼Hydrologic Budget (2)	Problem 4.18￼Unconfined Aquifer

Problem 4.19 Weep Holes Flow	Problem 4.20 Confined Aquifer (1)	📖 Problem 5.17 Adsorption using Powdered Activated Carbon PAC	Problem 5.18 Required Air Flow for Air Stripping
Problem 4.21 Confined Aquifer (2)	Problem 4.22 Composite Permeability	📖 Problem 5.19 Noncarbonate Hardness	Problem 5.20 Sedimentation Efficiency
Problem 4.23 Aquifer Porosity	Problem 4.24 Aquifer Volume	Problem 5.21 Water Distribution Reservoir	Problem 5.22 Population Growth

Analysis & Design

		📖 Problem 5.23 Concentration Gradient	📖 Problem 5.24 Depth of a Filter Bed
Problem 5.1 Influent Disinfection	Problem 5.2 Stormwater Drain Mixing	📖 Problem 5.25 Disinfection	📖 Problem 5.26 Giardia Cysts 3-LOG Inactivation by Free Chlorine
Problem 5.3 Lake Pollutant Decay	Problem 5.4 Water Treatment Plant Coagulation	Problem 5.27 Minimum Chlorine Residual	Problem 5.28 Short Term Lagoon Sizing
Problem 5.5 Inline Equalization Tank Volume Calculation	Problem 5.6 Primary Settling Tank Hydraulic Loading	📖 Problem 5.29 Aeration Basin Dissolved Oxygen Concentration	Problem 5.30 Equalization Tank Aeration Requirements
Problem 5.7 Offline Equalization Tank Volume Calculation	Problem 5.8 Rapid Filter Loading Rate	📖 Problem 5.31 Wastewater Treatment Process Selection	Problem 5.32 Chlorination Injection into the RAS Line
Problem 5.9 Organic Loading	Problem 5.10 Increase Depth of a Flow Using of a Weir	Problem 5.33 Biological Denitrification by adding Methanol	Problem 5.34 Dry Well Ventilation Requirements
Problem 5.11 Parshall Flume Design	Problem 5.12 Ogee Weir Design	Problem 5.35 Cascade Aeration System Design	(✻) Problem 5.36 Primary Anaerobic Sludge Digester
Problem 5.13 Design of a Primary Settling Tank	Problem 5.14 V-Notch Weir	Problem 5.37 Primary Settling Tank Scour Velocity	Problem 5.38 Trickling Filter BOD Removal
Problem 5.15 Rapid Mixing	(✻) Problem 5.16 Flocculator Design		

Problem 5.39 Activated Sludge MLSS Thickening	📖 Problem 5.40 Biological Denitrification /Biological Phosphorus Removal
Problem 5.41 Food to Microorganism Ratio	Problem 5.42 Returned Activated Sludge Flow (RAS)
Problem 5.43 Dechlorination	📖 Problem 5.44 Pond Water Quality
Problem 5.45 Stream Degradation caused by an Outfall	Problem 5.46 Oxygen Deficit caused by an Outfall
📖 Problem 5.47 Maximum Specific Growth Rate for Bacillus Subtilis	📖 Problem 5.48 Drinking Water Equivalent Level (DWEL)
📖 Problem 5.49 Probability of Risk from Carcinogenic Contaminant	Problem 5.50 Total Maximum Daily Loads TMDLs

Project Site Work

📖 Problem 6.1 Mass Haul Diagram	Problem 6.2 Volume of Excavation
Problem 6.3 Ground Level Uphill	Problem 6.4 Soil Loss Prevention
Problem 6.5 Soil Erodibility	Problem 6.6 Safety Incidence Rate
Problem 6.7 Excavation Construction Safety	Problem 6.8 Construction Activities
Problem 6.9 Construction Operation Over a Single Footing	Problem 6.10 Drainage Channel Cross Section
Problem 6.11 Basic Horizontal Curve	Problem 6.12 Basic Vertical Curve
Problem 6.13 Points on Vertical Curve	Problem 6.14 Crest Curve Slope
Problem 6.15 Lowest Point on Sag Curve	(✻) Problem 6.16 Retaining Wall Safety Factor
Problem 6.17 Retaining Wall Applicable Loads	Problem 6.18 Construction Methods
Problem 6.19 Balancing a Free Body	📖 Problem 6.20 Subbase Stabilization

(✻) Questions flagged like this either exceed typical exam length, or could be slightly more difficult than their counterparts, but they contain crucial concepts that are worth exploring. Those questions may also contain two concepts and are worth two questions not one. We decided to combine them in one question to deliver a certain concept, idea, or area of knowledge. We highly recommend that you practice all questions regardless of level of difficulty or length. Overall, in this guide, we reduced the number of simple questions and focused on the slightly more challenging ones to improve understanding.

📖 When you see this symbol, it indicates that additional theory and explanations are provided in the solution section for your convenience. While you are welcomed to skip it, we recommend taking a moment to explore the provided content.

PROBLEMS & SOLUTIONS

I. Project Planning

II. Soil Mechanics & Materials

III. Hydraulics

IV. Hydrology, Surface Water, Groundwater & Wells

V. Analysis & Design: Water, Wastewater & Water Quality

VI. Project Site Work

Theory Explained

Keep an Eye Out for This Symbol 📖

When you encounter this symbol in this book, it signifies that additional theory and explanation are provided in the solution section for your convenience.

While you are free to skip it, considering the wealth of knowledge related to water, wastewater, and environmental disciplines makes it advisable to stay informed.

Level of Difficulty (⁂)

In the following sections, few questions my exceed typical exam length or may require multiple values for an answer. You may encounter similar length or level of difficulty in your exam session but not all questions are going to be that long or difficult. However, we have elected to design this book for an ultimate experience for the PE exam and we did not want to leave anything for chance.

Difficult/ long questions are flagged with the symbol (⁂). We encourage you to attempt all questions regardless of their length or level of difficulty due the material they attempt to deliver.

I
PROJECT PLANNING

Knowledge Areas Covered

SN	Knowledge Area
1	**Project Planning**

    A. Quantity take-off methods
    B. Cost estimating
    C. Project schedules
    D. Activity identification and sequencing
    E. Economic and sustainability analysis (e.g., present worth, lifecycle costs, comparison of alternatives)

PART I
Project Planning

PROBLEM 1.1 *Project Cost Analysis*
The below table summarizes four activities which belong to a larger program:

Activity	Predecessor	Budget (USD)	Actual Cost (USD)	Progress
(A) Excavate pit	-	10,000	4,850	85%
(B) Compact & prep pit to receive precast units	(A)	2,000	-	0%
(C) Pour, cure, prep & transport precast units	-	18,000	8,500	60%
(D) Installation and backfilling	(C)	10,000	-	0%
	100%	40,000	13,350	

Activity	Wk1	Wk2	Wk3	Wk4	Wk5	Wk6	Wk7	Wk8
A	██	██						
B			██					
C			██	██	██	██	██	
D								██

At the end of week 3 the following statement best describes the project progress:

(A) Project is ahead of schedule and the estimate at completion is expected to be $27,668.

(B) Project is ahead of schedule and the estimate at completion is expected to be $34,050.

(C) Project is delayed as excavation works are behind schedule, based on cost performance, the project is expected to finish with a total cost of $27,668.

(D) Project is delayed as excavation works are behind schedule, based on cost performance, the project is expected to finish with a total cost of $34,050.

PROBLEM 1.2 *Project Activity Sequencing*
The below table represents project activities, durations in days, activity predecessors, and successors:

Activity	Duration	Predecessor	Successor
A	2	-	B, C
B	5	A	F, G
C	2	A	E, D
D	4	C	H
E	2	C	G
F	7	B	I
G	3	B, E	I
H	2	D	I
I	2	F, G, H	-

Assume work starts on a Monday, ignoring weekends and holidays, Total Float and Free Float for activity 'G' are:

(A) 5, 1

(B) 4, 4

(C) 4, 0

(D) 3, 3

PROBLEM 1.3 *Budgetary Cost Acceptable Range*
The city is conceptualizing a large infrastructure project with an expected cost of nearly $25 million. The city appointed a consultant to develop the feasibility and determine the budget for authorization from city council.

Assuming the project would in fact cost the city $25 million upon completion, what is an acceptable range of a budgetary estimate the consultant can provide the city with?

(A) $17.5 million to $37.5 million

(B) $20.0 million to $32.5 million

(C) $21.25 million to $30.0 million

(D) $20.0 million to $30.0 million

PROBLEM 1.4 *Benefit Cost Analysis*

The following path is $5\ miles$ long with $146 \times 10^6\ vehicle\ miles$ traveled on it per year:

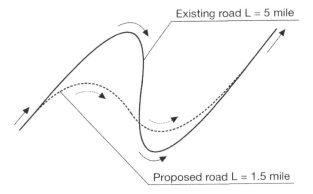

The dashed line is a proposed shortening of this segment which will cost $120.0\ million$ to erect and is 3.5 miles shorter (*).

Considering the following factors:

- Average fuel cost along with operation & maintenance cost is $ 0.14 per vehicle per mile.
- Road maintenance cost is $22,000 per mile per year.
- Inflation rate is estimated at 2%

The Benefit/Cost ratio for this project over a period of 25 years is most nearly:

(A) 0.91

(B) 1.17

(C) 2.33

(D) 1.31

(*) Prorate values to determine the total miles travelled on the shorter path.

(✻) PROBLEM 1.5 *Net Present Value Analysis*

One of the city's pumping stations require major mechanical rehabilitation for its pumps. The city has four options:

Option A: Replace old pumps with new pumps at a cost of $25,000 inclusive of labor. Yearly maintenance costs $3,000. The selected pumps to be replaced every 10 years.

Option B: Replace old pumps with new pumps at a cost of $27,500 inclusive of labor. Yearly maintenance costs $2,000. The selected pumps to be replaced every 10 years.

Option C: Replace old pumps with new pumps at a cost of $40,000 inclusive of labor. Yearly maintenance costs $1,500. The selected pumps to be replaced every 15 years.

Option D: Replace old pumps with new pumps at a cost of $47,500 inclusive of labor. Yearly maintenance costs $700. The selected pumps to be replaced every 15 years.

The remaining life of the civil structure is 30 years.

Average yearly inflation is 5% applied to capital and maintenance costs.

The city uses a discounted rate of 7% for their capital projects.

Based on the above, the cheapest and more feasible option amongst the four above would be:

(A) Option A

(B) Option B

(C) Option C

(D) Option D

(✻) This question exceeds typical exam length, but it contains crucial concepts that are worth exploring.

PROBLEM 1.6 *Resources Histogram*

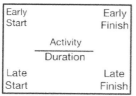

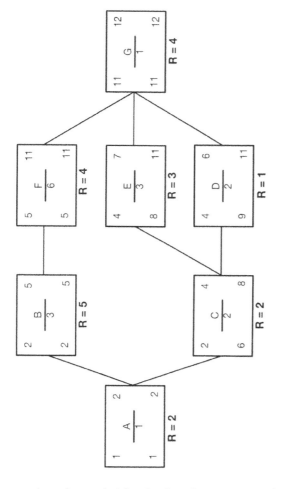

Assuming the activities in the above network diagram cannot be split, resources are of the same discipline, the below resources histogram represents the best possible leveled histogram (*):

(*) Activities start at the beginning of a day and finish at an end of a day – e.g., 'G' starts at the beginning of day 11 and ends at the end of day 11.

(A) Leveled Histogram A

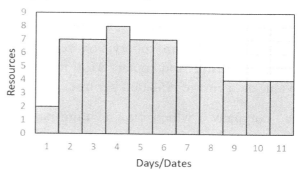

(B) Leveled Histogram B

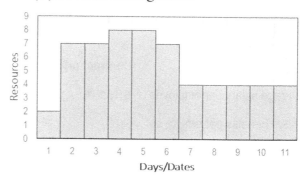

(C) Leveled Histogram C

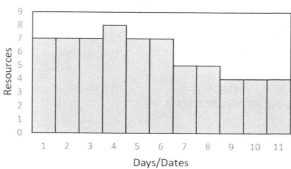

(D) Leveled Histogram D

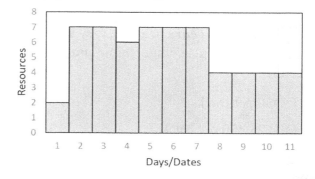

PROBLEM 1.7 *Salvage Value*

A company procured a fleet of high end, specially customized, old trucks for *five* of its employees to perform supervision services for a road construction contact $100 million in value at a cost of $55,000 per truck.

The company, which is a third-party contractor, priced its supervision services for this contract at 2.5% of the road construction fees.

The below is some of the key assumptions made by the company to support this agreement:

Item	Contract price
Average salary per employee	$80,000 / year
Fringe benefits per employee	30%
Overhead cost	12.5% of company revenue
Allocations	7.5% of company revenue
Fuel & maintenance per truck	$0.25 / mile
Expected miles driven for supervision purposes	95 mile/day /employee
Working days per year (excluding vacation time)	240 days
Contract duration	2.5 years

In order to maintain an Operating Income OI of 15%, and considering those trucks are pretty much worn out after being driven over rough land for 2.5 *years* in a row, they should be salvaged at an average minimum sales price of:

(A) $0 per truck

(B) $21,250 per truck

(C) $4,250 per truck

(D) $9,250 per truck

PROBLEM 1.8 *Profit and Loss*

You are a formwork/wood supplier who has been asked to provide material for one of the contractors to erect a building project. The material will cost you $100,000 to supply. The contractor asked you for a monthly payment plan over one year for cash flow reasons.

The monthly charge for the contractor that gets you to maintain an overall profit margin of 20% on all your expenses, knowing that bank charges a fixed interest of 6% per year, is most nearly:

(A) $10,600

(B) $10,952

(C) $10,332

(D) $10,720

PROBLEM 1.9 *Capitalized Costs*

The present net worth at a 4% interest rate for a project that has an initial cost of $1,000,000 and an Operation and Maintenance cost of $75,000 is most nearly:

(A) $2,875,000

(B) $1,875,000

(C) $1,750,000

(D) $2,171,650

PROBLEM 1.10 *Rate of Concrete Pouring*

The below top section is a plan view for a 14 *in* thick 11 *ft* high wall combination prepared for concrete placement for a certain facility.

The facility under consideration has eight of those wall combinations and concrete should be placed in sequence starting from wall combination 1 to wall combination 8.

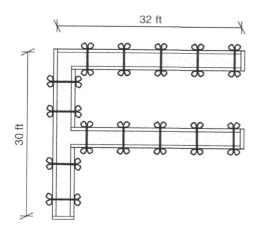

Concrete placement starts at 8:00 AM and must stop at 12:00 PM. Assuming concrete is pumped continuously at an average rate of 70 $yard^3/hr$ inclusive of lost time, a horizontal construction joint, if needed during the four-hour window, will be placed at:

(A) Wall combination 6 at a height of 0.5 ft

(B) Wall combination 6 at a height of 5 ft

(C) Wall combination 7 at a height of 5 ft

(D) Wall combination 7 at a height of 2.5 ft

PROBLEM 1.11 *Bond Effective Interest Rate*

A bond worth $200,000 that pays 7% yearly compounded interest realized every quarter for a year has an effective annual interest rate of:

(A) 7.2%

(B) 7.0%

(C) 7.5%

(D) 31%

PROBLEM 1.12 *Depreciated Rate*

A machinery was bought by a construction company for $125,000 and it was estimated that it would be salvaged in 10 years for $10,000 which represents the end of its life.

The rate of depreciation for this machinery is most nearly:

(A) 9%

(B) 10%

(C) 11%

(D) 12%

SOLUTION 1.1

Reference is made to the earned value management equations in *NCEES Handbook* based upon which the below tables are constructed:

Activity	I Duration (Weeks)	Budget (USD)	II Actual Cost (USD)	III Progress
A	2.0	10,000	4,850	85%
B	1.0	2,000	-	0%
C	6.0	18,000	8,500	60%
D	1.0	10,000	-	0%
	8.0	40,000	13,350	

IV Planned completion wk3	I × IV BCWS Planned Budget (USD)	II ACWP Actual Cost (USD)	I × III BCWP Earned Value (USD)
100%	10,000	4,850	8,500
100%	2,000	-	-
33.3%	6,000	8,500	10,800
0	-	-	-
	18,000	13,350	19,300

$CPI = BCWP/ACWP$
$= 19,300/13,350$
$= 1.445$

$SPI = BCWP/BCWS$
$= 19,300/18,000$
$= 1.07$

$ETC = (BAC - BCWP)/CPI$
$= (40,000 - 19,300)/1.445$
$= \$ 14,318$

$EAC = ACWP + ETC$
$= \$13,350 + \$14,318$
$= \$ 27,668$

The above indicates that the project is ahead of schedule with SPI >1.0 and cost performance is positive with CPI >1.0 and the estimate to completed is $27,668.

Correct Answer is (A)

SOLUTION 1.2

Using the critical path method, start with building an activity network. *Activity on nodes* was the preference in this solution with the following nomenclature and further steps:

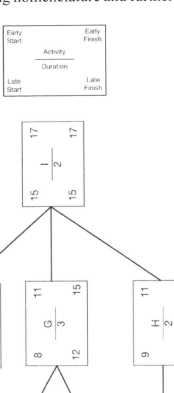

Step 1: A *forward pass* to determine early dates (Early Start ES and Early finish EF). When two activities feed forward into the same successor, the latest (longest) EF prevails.

Step 2: A *backward pass* to determine late dates (Late Start LS and Late finish LF). When two activities feed backward into the same predecessor, the earliest (shortest) LF prevails.

The following equations were applied to activity 'G':

$Total\ Float = LS - ES = 12 - 8 = 4\ days$

$Free\ Float = Earliest\ ES_{successor} - EF$
$= 15 - 11 = 4\ days$

Correct Answer is (B)

SOLUTION 1.3
This is a common request in real projects and is usually backed up with principles from the *Association for the Advancement of Cost Engineering* AACE.

The *Cost Estimate Classification Matrix* adopted by the AACE is found in the *NCEES Handbook version 2.0* Section 2.2.2. This matrix classifies estimates for Design Development, Budget Authorization and Feasibility as Class 3 Estimates.

The expected accuracy range for an AACE Class 3 estimate is:

L: − 5% to − 15%
(i.e., **$21.25** *million* to $23.75 *million*)

H: + 10% to + 20%
(i.e., $27.5 *million* to **$30.0 million**)

The range would therefore be constituted from the lowest which is $21.25 *million* to the highest being $30.0 *million*.

Correct Answer is (C)

SOLUTION 1.4
To calculate the Benefit/Cost ratio for the path shortening project, we need to determine the total benefits and total costs over the 25 years period.

Costs:
Annual operation & maintenance and fuel costs are brought to their Present Values (PV) using Table 1.7.10 of the *NCEES Handbook version 2.0* with an interest rate of $i = 2\%$ and $n = 25\ years$ represented as $(P/A, 2\%, 25) = 19.5$. Cost is then calculated as follows:

(A) Annual cost before shortening is calculated as follow:

Road maintenance cost:
$= 5\ miles \times \$\ 22,000 \times 19.5$
$= \$\ 2,145,000\ (\$\ 2.145\ million)$

The estimate for the annual vehicle miles traveled for the original path length of $5\ miles$ is given in the question as:

$146 \times 10^6\ miles\ per\ year$

Vehicle fuel and O&M cost:
$= 146 \times 10^6\ miles \times \$\ 0.14 \times 19.5$
$= \$\ 398.58\ million$

Total fuel and vehicle O&M cost $(TC_{O\&M\ before})$:
$= \$\ 398.58\ million + \$\ 2.145\ million$
$= \$\ 400.73\ million$

(B) Annual cost after improvements is calculated as follow:

Road maintenance cost:
= 1.5 miles × $ 22,000 × 19.5
= $ 643,500 ($ 0.64 million)

The new estimate for the annual vehicle miles traveled for the new path length of 1.5 miles is determined as suggested in the question as follows:

$$\frac{1.5}{5} \times 146 \times 10^6$$
$$= 43.8 \times 10^6 \text{ miles per year}$$

Vehicle fuel and O&M cost:
= 43.8 × 10⁶ miles × $ 0.14 × 19.5
= $ 119.574 million

Total fuel and vehicle O&M cost ($TC_{O\&M\ after}$):
= $ 119.57 million + $ 0.64 million
= $ 120.21 million

Total Cost after improvement (TC_{after}):
= $ 120.21 million + $ 120.00 million
= $ 240.21 million

<u>Benefits:</u>
The benefits arise from the cost savings due to reduced fuel and O&M for vehicles, along with the maintenance cost for the improved road segment over the specified duration of 25 years.

$$= TC_{O\&M\ before} - TC_{O\&M\ after}$$
$$= \$\ 400.73\ million - \$\ 120.21\ million$$
$$= \$\ 280.52\ million$$

<u>Benefit/Cost ratio:</u>

B/C is calculated by dividing the total benefits (savings) by the total cost:

$$B/C = \frac{Savings}{Total\ Cost\ (TC_{after})}$$

$$= \frac{\$\ 280.52\ million}{\$\ 240.21\ million}$$

$$= 1.17$$

A B/C of 1.17 suggests that for every dollar invested in the path shortening, you would receive approximately $ 1.17 in benefits over the 25 years period. Typically, a B/C greater than '1.0' is considered economically viable, as it indicates that the benefits outweigh the costs (*).

Correct Answer is (B)

(*) It is essential to consider other factors in the Benefit Cost analysis such as potential future savings, environmental impacts, improved safety, and additional intangible benefits that may not be captured in this simple analysis. A more comprehensive evaluation in real life examples may be necessary to make well-informed decisions.

📖 (✻) SOLUTION 1.5

Although this seems a slightly longer question than an exam question would be, it provides an explanation of the concept of NPV and options analysis for a real, and frequent, engineering problem.

One of the means of performing options analysis is the *Net Present Value* (NPV) method. With this, future costs shall be determined with using average inflation. After that, NPV is calculated per option using the given *discounted rate* as *interest rate*.

Step 1: Calculate inflation and inflation factors as follows:

Step 2: Perform NPV analysis as follows:

Option A

Event	Year	n	Cost (I)	Inflation Symbol	Inflation Factor (II)	Inflated Cost I x II	NPV Symbol	NPV Factor	NPV
New pumps	0		$25,000	-	1.0000	$25,000.00	NA	1.0000	$25,000.00
New pumps	10	10	$25,000	(F/P, 5%, 10)	1.6289	$40,722.50	(P/F, 7%, 10)	0.5083	$20,699.25
New pumps	20	20	$25,000	(F/P, 5%, 20)	2.6533	$66,332.50	(P/F, 7%, 20)	0.2584	$17,140.32
Maintenance	1→9 & 10→19 & 20→30	28	$3,000	(F/A, 5%, 30) minus 2 x (F/P, 5%, 30)	66.4388 minus 2 x 4.3219	$173,385.00	(P/F, 7%, 30)	0.1314	$22,782.79
									$85,622.35

Option B

Event	Year	n	Cost (I)	Inflation Symbol	Inflation Factor (II)	Inflated Cost I x II	NPV Symbol	NPV Factor	NPV
New pumps	0		$27,500	-	1.0000	$27,500.00	NA	1.0000	$27,500.00
New pumps	10	10	$27,500	(F/P, 5%, 10)	1.6289	$44,794.75	(P/F, 7%, 10)	0.5083	$22,769.17
New pumps	20	20	$27,500	(F/P, 5%, 20)	2.6533	$72,965.75	(P/F, 7%, 20)	0.2584	$18,854.35
Maintenance	1→9 & 10→19 & 20→30	28	$2,000	(F/A, 5%, 30) minus 2 x (F/P, 5%, 30)	66.4388 minus 2 x 4.3219	$115,950.00	(P/F, 7%, 30)	0.1314	$15,235.83
									$84,359.35

Option C

Event	Year	n	Cost (I)	Inflation Symbol	Inflation Factor (II)	Inflated Cost I x II	NPV Symbol	NPV Factor	NPV
New pumps	0		$40,000	-	1.0000	$40,000.00	NA	1.0000	$40,000.00
New pumps	15	15	$40,000	(F/P, 5%, 15)	2.0789	$83,156.00	(P/F, 7%, 15)	0.3624	$30,135.73
Maintenance	1→14 & 15→30	29	$1,500	(F/A, 5%, 30) minus 1 x (F/P, 5%, 30)	66.4388 minus 1 x 4.3219	$93,175.35	(P/F, 7%, 30)	0.1314	$12,243.24
									$82,378.98

Option D

Event	Year	n	Cost (I)	Inflation Symbol	Inflation Factor (II)	Inflated Cost I x II	NPV Symbol	NPV Factor	NPV
New pumps	0		$47,500	-	1.0000	$47,500.00	NA	1.0000	$47,500.00
New pumps	15	15	$47,500	(F/P, 5%, 15)	2.0789	$98,747.75	(P/F, 7%, 15)	0.3624	$35,786.18
Maintenance	1→14 & 15→30	29	$700	(F/A, 5%, 30) minus 1 x (F/P, 5%, 30)	66.4388 minus 1 x 4.3219	$43,481.83	(P/F, 7%, 30)	0.1314	$5,713.51
									$88,999.70

Option (A) explained:
- Year zero: new pumps to be procured at a cost of $25k$. The $25k$ is the present value of this initial engineering decision and hence no factor is applied on it and this cost shall be taken to the finish line (i.e., NPV) as is.

- Year 10: new pumps to be procured at a cost of $25k$ *now*. In 10 years however, the cost of those pumps is expected to be $(F/P, 5\%, 10)$ more – i.e.,
$25,000 \times 1.6289 = \$40,722.5$

The $40,722.5 is brought back to year zero using the city discounted rate (i.e., the best estimate for their return on investment on capital projects) using $(P/F, 7\%, 10)$ – i.e.,
$40,722.5 \times 0.5083 = \$20,699.25$

- Year 20: new pumps to be procured at a cost of $25k$ *now* as given in the question. In 20 years however, the cost of those pumps is expected to be $(F/P, 5\%, 20)$ more – i.e., $66,332.5$. This amount is brought back to year zero using $(P/F, 7\%, 20)$ as:
$(25,000 \times 2.6533) \times 0.2584 = 17,140.3$

- Years 1 to 30: yearly maintenance cost of $3,000. With inflation, this cost is brought forward to year 30 in one lump with the factor $(F/A, 5\%, 30)$ as:
$3,000 \times 66.4388 = \$199,316.4$

There are two years however during when maintenance costs are not required which is when pumps are installed at year 10 and at year 20. Those two years' payments are deducted in year 30 with $2 \times (F/P, 5\%, 30)$ as:
$2 \times 4.3219 \times \$3,000 = \$25,931.4$

→ $199,319.4 - \$25,931.4 = \$173,385.0$

The overall net maintenance resultant in year 30 is therefore expected to be $173,385.0. This amount is discounted (i.e., brought back to year zero) with the factor $(P/F, 7\%, 30)$ to:
$173,385.0 \times 0.1314 = \$22,782.79$

The total NPV for option A in this case is:

$25,000 + \$20,699.25 + \$17,140.32 + \$22,782.79 = \$85,622.36$

A similar process is applied to all the given options and option C was the most feasible.

Correct Answer is (C)

SOLUTION 1.6

It is important to understand which activities can change their start and finish dates so they can be used for resources leveling. Those activities are the ones with positive free float. Free float can be determined using the following equation:

$Free\ Float = Earliest\ ES_{successor} - EF$

$FF_D = 11 - 6 = 5\ days$

$FF_E = 11 - 7 = 4\ days$

The following table summarizes the network diagram along with the available Free Floats:

Activity	Duration	Resources	ES	EF	FF
A	1	2	1	2	
B	3	5	2	5	
C	2	2	2	4	
D	2	1	4	6	5
E	3	3	4	7	4
F	6	4	5	11	
G	1	4	11	12	

This table is then converted into the base Gantt Chart shown below, out of which a resources histogram is drawn by assigning resources to the bar chart and summing them up in the histogram that follows.

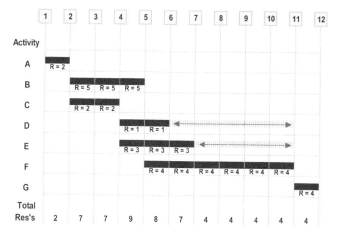

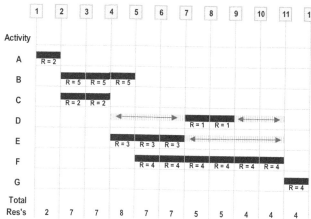

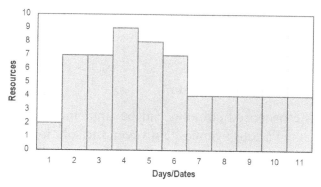

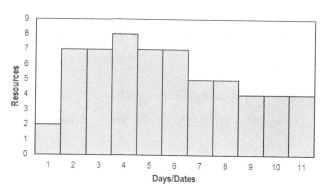

As pointed out in the above Gantt Chart, activities D and E can move freely as indicated by the two-sided arrows for *five* and *four* days respectively without affecting the completion date of the project.

In this case, the solution becomes a matter of trial and error. Activities D and E can move horizontally within their float to match the best available option from the four options provided in this question.

Based on this, the following Gantt chart was established by sliding activity D *three* days to the right which could provide the most leveled usage for those resources, or at least to match one of the histograms provided in the question.

The rest of the diagrams are simply incorrect.

Correct Answer is (A)

📖 SOLUTION 1.7

Company revenue:

Contract fees (company revenue)

$= \$100,000,000 \times 2.5\%$

$= \$2,500,000$

Project cost and Gross Margin (GM):

Direct labor (salaries)

$= \$80,000 \times 5 \times 1.3 \times 2.5$

$= \$1,300,000$

Truck expenses

$= \$0.25 \times 5 \times 95 \; mile \times 240 \times 2.5$

= $71,250

Total project expenses

= $1,300,000 + $71,250

= $1,371,250

Gross Margin (GM)

= $2,500,000 − $1,371,250

= $1,128,750

Gross Margin (%)

= $1,128,750/$2,500,000

= 45.2%

Company Operating Income (OI):

$OI = GM - OH - Capital\ Cost - Alloc$

Overhead cost (OH)

$= Revenue \times OH\%$

= $2,500,000 × 12.5%

= $312,500

Company capital costs (trucks in this case)

= $55,000 × 5

= $275,000

Allocations (Alloc)

$= Revenue \times Alloc\%$

= $2,500,000 × 7.5%

= $187,500

$OI = \$1,128,750 - \$312,500 - \$275,000 - \$187,500$

= $353,750

$OI\ (\%) = \$353,750/\$2,500,000$

= 14.2%

The capital value for trucks is not usually factored into projects' Gross Margins, this is why this value was taken towards the bottom line.

Therefore, to maintain an Operating Income of :

$OI = \$2,500,000 \times 15\% = \$375,000$

Trucks should be salvaged at a minimum of:

$\$375,000 - \$353,750 = \$21,250$

$\$21,250/5 = \$4,250\ per\ truck$

Correct Answer is (C)

📖 Solution discussion:

The proposed solution can be implemented using various avenues. However, the solution was presented in a manner consistent with the prevalent practices observed among professional organizations for their financial reporting.

In this context, company revenue (or part of it) is represented by its contract fees of $2,500,000.

The overhead cost $312,500 includes expenditures related to marketing, administration, and other non-chargeable project costs, and is customarily computed. It is reasonable to consider this cost as a percentage of the company's overall revenue, in this case 12.5%.

The same principle applies to allocations calculated as $187,500. In certain instances, particularly within large organizations, these costs may be separately itemized, especially in cases where the company operates across multiple regions, maintains various offices, and supports executive functions such as the C-suite and the President's office.

Direct labor accounts for actual staff salaries which averages at $80,000, while the 30% fringe benefits incorporate provisions for annual leave, medical leave, and other contractual benefits granted to employees.

Although trucks' capital cost was not factored into the project cost, it is noteworthy that such capital expenditures reflect a deliberate choice made by the company when alternate options, such as renting trucks, could be available. However, the assumption to include trucks capital cost within the capital expenditure framework has been made as part of the solution. Consequently, such costs do not factor into the project margin calculation.

In other instances, the company would (internally) rent out those trucks (after procuring them) to the project and reduces its project Gross Margin. None of these methods would change the final outcome of the solution.

Truck expenses (such as mileage and operation and maintenance costs = $71,250 over 2.5 *years*) are classified as project-related expenditures, and therefore reduces project's Gross Margin or its profitability.

Operating Income OI in this case $353,750 – which, is also termed as EBITDA (Earnings Before Interest, Taxes, Depreciation, and Amortization) – is derived by subtracting the overhead costs $312,500, allocations $187,500, and any other associated capital investments (exemplified by trucks in this case) $275,000, from the project's Gross Margin $1,128,750.

Considering the scenario at hand, where the company did not achieve the target Operating Income from this project $375,000, selling those trucks and realizing a minimum final sales value of $4,250 per truck could be deemed a viable option, particularly if there are no alternative contracts that necessitate the use of the trucks and given those trucks are at the end of their operating life.

SOLUTION 1.8
Project expenses include the cost of material along with the cost of borrowing, in which case you will be the one borrowing from the bank on behalf of the contractor to supply them with the material.

$$Cost\ of\ borrowing = 6\% \times \$100,000$$
$$= \$6,000$$

$$Cost\ of\ material = \$100,000$$

$$Total\ expenses = \$100,000 + \$6,000$$
$$= \$106,000$$

$$Overall\ sales\ inclusive\ of\ profit\ on\ all\ expeses$$
$$= \$106,000 \times 1.2$$
$$= \$127,200$$

$$Monthly\ payment = \frac{127,200}{12} = \$10,600$$

Correct Answer is (A)

SOLUTION 1.9
The present net worth is the capitalized cost (P) for a constant annual cost over an infinite period. See Capitalized costs equation in Section 1.7.7 of the *NCEES Handbook* version 2.0.

$$P = \$1,000,000 + \frac{75,000}{4\%} = \$2,875,000$$

Correct Answer is (A)

SOLUTION 1.10
There are few methods that can be used to solve this question, below is one.

Start with calculating the surface area for one wall combinations as follows:

$$A = 2 \times \left[\left(32 - \frac{14}{12}\right) \times \frac{14}{12}\right] + 30 \times \frac{14}{12}$$
$$= 106.9 \, ft^2$$

Rate of concrete placement in ft^3/hr:

$$\alpha = 70 \, yard^3/hr \times \frac{27 \, ft^3}{yard^3}$$
$$= 1,890 \, ft^3/hr$$

In *four* hours, the volume of pumped concrete should be as follows:

$$V = 1,890 \, ft^3/hr \times 4 \, hrs$$
$$= 7,560 \, ft^3$$

Height of concrete poured in *four* hours:

$$h = \frac{7,560 \, ft^3}{106.9 \, ft^2} = 70.7 \, ft$$

Divide this by each wall combination height to determine how many walls could be poured in full:

$$No. of \, walls = \frac{70.7 \, ft}{11 \, ft} = 6.43 \, wall$$

This answer represents *six* walls fully poured and 43% of the seventh wall being poured which equals to 43% × 11 = 4.73 ft

Correct Answer is (C)

SOLUTION 1.11
The effective annual interest rate is the adjusted annual rate after compounding over a given period. See Nonannual Compounding and effective interest rate equation in Section 1.7.2 of the *NCEES Handbook version 2.0*.

$$i_e = \left(1 + \frac{r}{m}\right)^m - 1 = \left(1 + \frac{7\%}{4}\right)^4 - 1 = 7.19\%$$

Correct Answer is (A)

SOLUTION 1.12
The depreciation rate is calculated using annual depreciation (straight line method can be used) divided by the depreciable cost.

Using *NCEES Handbook version 2.0* Section 1.7.5, straight line depreciation is computed as follows:

$$D_j = \frac{C - S_n}{n}$$
$$= \frac{\$125,000 - \$10,000}{10}$$
$$= \$11,500$$

$$Rate \, of \, depreciation = \frac{\$11,500}{\$125,000 - \$10,000}$$
$$= 10\%$$

Correct Answer is (B)

II
SOIL MECHANICS & MATERIALS

Knowledge Areas Covered

SN	Knowledge Area
2	**Soil Mechanics**
	A. Lateral earth pressure
	B. Soil consolidation and compaction
	C. Bearing capacity
	D. Settlement
	E. Slope stability
3	**Materials**
	A. Soil classification and boring log interpretation
	B. Soil properties (e.g., strength, permeability, compressibility, phase relationships)
	C. Concrete (e.g., nonreinforced, reinforced)
	D. Piping materials
	E. Material test methods and specification conformance

PART II
Soil Mechanics & Materials

PROBLEM 2.1 *Pressure Under Footing*

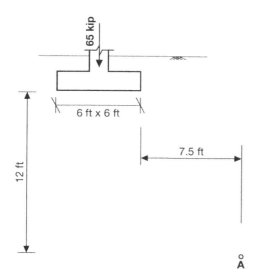

The increase in the vertical pressure at point 'A' due to loading the square footing using those two theories respectively:

- The Boussinesq's Theory
- The 2:1 theory

are as follows:

(A) ≈ 60 psf (Boussinesq) and zero (2:1 theory)

(B) ≈ 60 psf (Boussinesq) and 200 psf (2:1 theory)

(C) ≈ 200 psf (Boussinesq) and zero (2:1 theory)

(D) ≈ 200 psf (Boussinesq) and 200 psf (2:1 theory)

PROBLEM 2.2 *Foundation Settlement*

The below is a 6 ft × 6 ft square concrete footing laid in fine medium dense sand with a maximum load applied on it of 100 kip.

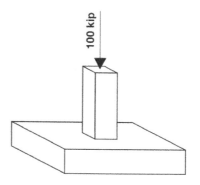

The maximum initial elastic vertical settlement this footing will experience is most nearly:

(A) 0.77 in

(B) 0.064 in

(C) 1.55 in

(D) 0.46 in

📖 PROBLEM 2.3 *Effective Stress Over Time*

The below matt foundation has a uniform weight of 1 ksf and was built and loaded linearly over a period of 6 months on top of a layer of sand and clay as shown. The ground water level is 5 ft below the footing. The density of sand and clay layers are both 120 pcf.

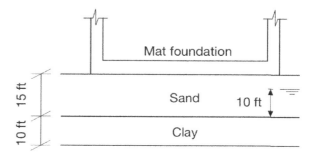

The profile that represents the change in effective stress over time at the bottom of each layer is:

(A) Profile A

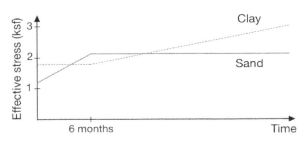

(B) Profile B

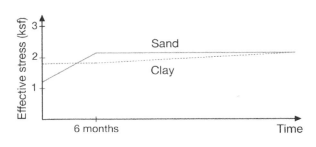

(C) Profile C

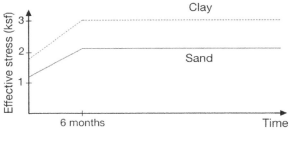

(D) Profile D

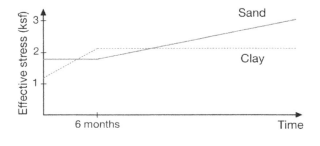

PROBLEM 2.4 *Optimum Moisture Content*

A proctor test was performed on *four* soil samples from the same batch using a proctor standard mold.

The weight of the moist samples after applying the test blows were: 4.32 lb, 3.92 lb, 3.92 lb and 4.36 lb, and their moisture content 11.6%, 17%, 8.8% and 14.8% respectively (*).

The optimum moisture content for this batch is most nearly:

(A) 14.0%

(B) 11.3%

(C) 12.5%

(D) 14.8%

(*) Take volume of mold as $1/30\ ft^3$

PROBLEM 2.5 *Slope Stability/ Slope Safety Factor*

The below slope belongs to an excavation in a soil with cohesion $c = 58\ psf$ and density $\gamma = 95\ pcf$ along with a friction angle of $\emptyset = 20°$.

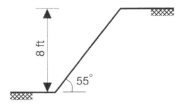

Using Taylor soil stability charts, the slope safety factor for this excavation is most nearly:

(A) 0.9

(B) 1.5

(C) 3.0

(D) 0.3

PROBLEM 2.6 Consolidation Settlement

The below graph plots the results of an odometer test for a clay sample where x-axis represents the logarithm of pressure, and y-axis is the void ratio (e).

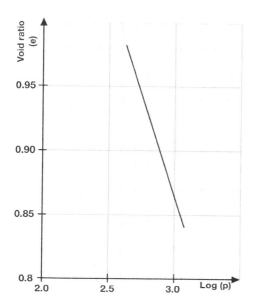

The expected settlement for a $4\ ft$ thick layer of this clay when pressure increases from an initial pressure of $500\ psf$ to a final pressure of $1,000\ psf$ is most nearly:

(A) $2.4\ in$

(B) $0.2\ in$

(C) $3.8\ in$

(D) $0.3\ in$

PROBLEM 2.7 Bearing Capacity for a Square Foundation

The below is a concrete square foundation placed $5\ ft$ below surface in soil that has the following properties:

- Cohesion = $450\ psf$
- Friction angle = $35°$
- Density = $130\ pcf$
- Ground water $5\ ft$ deep

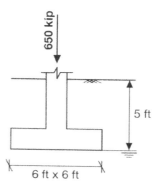

Considering the single footing carries a load of $650\ kip$, safety factor for this footing against shear failure is most nearly:

(A) 4.3

(B) 2.9

(C) 3.6

(D) 4.6

(✻) PROBLEM 2.8 Soil Properties

A soil sample that has a total volume of $1\ ft^3$ and a total mass of $100\ lb$ is removed from the ground. The water content of this sample is 20% and Specific Gravity '2.7'.

Based on the above information the following attributes are as follows:

The dry density of the sample is _____

The Degree of Saturation is _____

Porosity is _____

(✻) Normally you are requested to provide one value.

PROBLEM 2.9 Dam Site Rock Quality

Drilling was carried out for a dam site investigation project for a depth of $100\ ft$. Total length of recovered core pieces for samples $> 4\ in$ add up to $70\ ft$.

The below best describes the quality of rock:

(A) Very Poor

(B) Poor

(C) Fair

(D) Good

(✲) PROBLEM 2.10 Soil Properties (2)

The over consolidation ratio for a soil with a '0.33' normally consolidated at rest Rankine coefficient and a '0.85' over consolidated at rest coefficient is most nearly:

(A) 4.1

(B) 1.9

(C) 0.3

(D) 2.6

PROBLEM 2.11 Soil Shear Strength

The below is a particle distribution chart for two sands, Sand 1 (top) and Sand 2 (bottom).

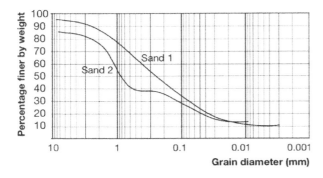

Based on this chart, the following statement is accurate during normal loading for the two sands:

(A) Sand 2 has a higher shear strength compared to Sand 1.

(B) Sand 1 has a higher shear strength compared to Sand 2.

(C) Both Sands have approximately the same shear strength.

(D) More information is required.

PROBLEM 2.12 Settlement in Clay

Time settlement in saturated clays when loaded, due to the addition of a building for example, is attributed to the following:

(A) Expulsion of clay particles

(B) The increase in effective stress of clay

(C) The deformation of clay particles

(D) All the above

PROBLEM 2.13 Boreholes Locations

The below is a cross section of a $300\ ft$ long, yet to be designed, retaining wall.

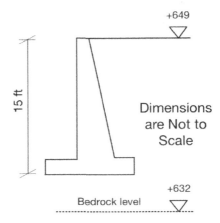

The minimum number, locations, and depth of the needed exploration boreholes to design this wall is/are:

(A) One borehole $6\ ft$ deep in front of the wall below its footing level. Another one $21\ ft$ (*) behind the wall.

(B) One borehole 16 ft deep in front of the wall below its footing level. Another one 31 ft (*) behind the wall.

(C) One borehole 21 ft (*) deep and another borehole 35 ft (*) deep both behind the wall.

(D) One borehole 35 ft (*) deep behind the wall at mid span.

(*) Take datum level as (+649) for boreholes taken behind the wall.

PROBLEM 2.14 *Soil Moisture Content*

The maximum moisture content of a soil is close to its:

(A) Liquid limit

(B) Shrinkage limit

(C) Plastic limit

(D) Plasticity index

PROBLEM 2.15 *Soil Classification System*

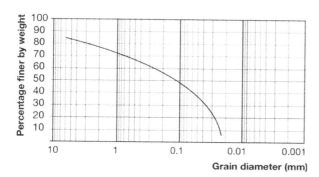

Based on the above soil sample gradation, and the following fines attributes:

- Liquid Limit (LL) = 50
- Plastic Limit (PL) = 35

Using the AASHTO classification system, the following represents the best group classification for this sample:

(A) A-7-5 (4)

(B) A-7-6 (4)

(C) A-7-4

(D) A-7-5

PROBLEM 2.16 *Soil Classification System*

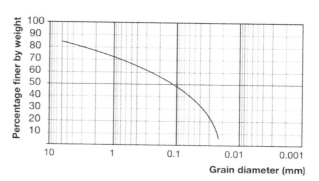

Using the Unified Soil Classification System, the above gradation sample is (*):

(A) GW

(B) GP

(C) SW

(D) SM

(*) Assume fines are classified as MH.

PROBLEM 2.17 *Concrete Mix Design*

A concrete mix with a specified w/c ratio of 0.45, a mix design of 1:1.5:3, assuming aggregate (*) is 5% in general moisture deficit to get into SSD, requires ____ liters of water to produce 1 ft^3 of yield.

(A) 5.4

(B) 6.5

(C) 10.5

(D) 19.3

(*) Density of aggregate and cement are 165 lb/ft^3 and 195 lb/ft^3 respectively.

PROBLEM 2.18 *Soil Permeability Testing*
The following test is recommended to be used to determine the coefficient of permeability for materials with lower permeability such as silts and clays:

(A) The constant head permeameter test

(B) The Piezometric test

(C) The falling head permeameter test

(D) The flexible wall permeameter test

PROBLEM 2.19 *Cement Type*
The best type of cement that is preferred for the use in a sewage treatment facility's deep thick base for one of its clarifiers is:

(A) Type I – Ordinary

(B) Type III – High early strength

(C) Type IV – Low heat

(D) Type V – Sulfate resistant

PROBLEM 2.20 *Hydrostatic Pressure Test*
Newly installed water ductile iron pipes are tested in accordance with the AWWA C600 standard by applying 1.25 times the operating pressure such that the change in pressure does not exceed 5 *psi* for ___ hour(s):

(A) 0.5

(B) 1

(C) 1.5

(D) 2

PROBLEM 2.21 *Pipe Material (1)*
A potable water pipeline is to be installed with little amount of supervision involved. The best material for use in this case is:

(A) Concrete pipe

(B) Fiberglass pipe

(C) HDPE pipe

(D) Ductile iron pipe

PROBLEM 2.22 *Pipe Material (2)*
The preferred pipe material to be used for ozone service in water treatment plants is:

(A) Low Carbon 304L

(B) Low Carbon 316L

(C) Polyethylene

(D) Sch80 seamless steel

SOLUTION 2.1

The pressure right below the footing is calculated as follows:

$$q_o = \frac{P}{A} = \frac{65}{6 \times 6} = 1.8 \text{ ksf}(= 1{,}800 \text{ psf})$$

Boussinesq's method:
Using the square footing part of Boussinesq's Isobars chart – copied below for ease of reference, the horizontal and vertical coordinates of the chart are determined as portions of B (i.e., the footing width) as follows:

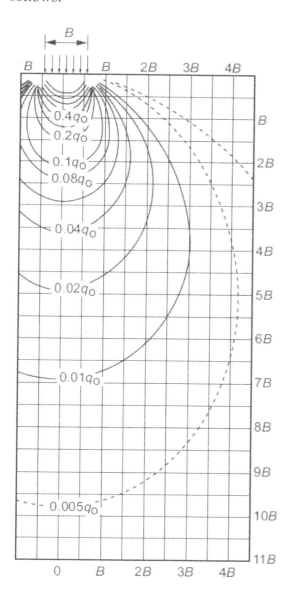

Interpolating those coordinates using the Isobar chart:

$$\Delta P = 0.035 q_o = 0.035 \times 1{,}800 = 63 \text{ psf}$$

The 2:1 method:
The 2:1 method assumes a 2:1 trapezoidal distribution of the load as shown in the following figure:

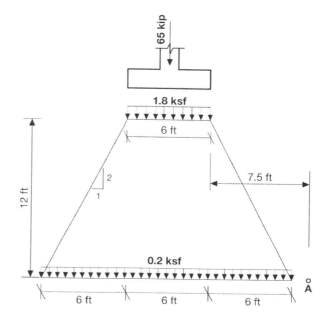

Based on the above distribution, a vertical depth of $Z = 12\text{ ft}$ corresponds to a horizontal distance from the face of the footing of 6 ft. Point 'A' however sits at 7.5 ft from the edge of the footing, i.e., 1.5 ft away.

The pressure at point 'A' using this method is therefore Zero.

Correct Answer is (A)

SOLUTION 2.2
Referring to the *NCEES Handbook version 2.0*, Section 3.5.2, the vertical elastic settlement is calculated as follows:

$$\delta_v = \frac{C_d \Delta p B_f (1 - v^2)}{E_m}$$

C_d is the rigidity factor and is looked up from the table in the same section as '0.99' for rigid (i.e., concrete) square shaped foundations. Δp is the increase in pressure right below the foundation which is ($100 kip/36 ft^2$). B_f is the footing dimension which is $6 ft$ for squared footings in this case.

For fine medium dense sand, Poisson's ratio v is '0.25' as collected from the same Section. The young modulus E_m shall be the lowest of the range provided in the guide – $120\ tsf$ in this case – as the question is looking for maximum settlement.

$$\delta_v = \frac{0.99 \times \left(\frac{100}{36}\right) kip/ft^2 \times 6ft \times (1 - 0.25^2)}{120 \frac{ton}{ft^2} \times 2 \frac{kip}{ton}}$$

$$= 0.064\ ft\ (0.77 in)$$

Correct Answer is (A)

📖 SOLUTION 2.3
The effective stress σ' is the total stress σ removed from it the pore/water pressure u. In which case, the following effective stress changes at the bottom of each layer shall occur over the indicated period of 6 months and the age of the project/embankment.

Month zero prior to placing the embankment:
$\sigma_{sand} = 120\ pcf \times 15\ ft$
$\qquad = 1,800\ psf\ (1.8\ ksf)$

$u_{sand} = 62.4\ pcf \times 10\ ft$
$\qquad = 624\ psf\ (0.62\ ksf)$

$\sigma'_{sand} = \sigma_{sand} - u_{sand}$
$\qquad = 1.8 - 0.62 \cong 1.2\ ksf$

$\sigma_{clay} = 120\ pcf \times (10 + 15)\ ft$
$\qquad = 3,000\ psf\ (3.0\ ksf)$

$u_{clay} = 62.4\ pcf \times 20\ ft$
$\qquad = 1,248\ psf\ (1.25\ ksf)$

$\sigma'_{clay} = \sigma_{clay} - u_{clay}$
$\qquad = 3.0 - 1.25 \cong 1.8\ ksf$

Month 6 after placing the embankment:
When the embankment is placed, the sand layer will drain the excess (now pressurized) water immediately and hence no increase shall occur in the sand pore pressure. This will reflect in an increase in the effective pressure of the sand and given the loading from the embankment took place linearly over a period of 6 months, the increase in effective pressure for sand will be linear as well.

$\sigma_{sand} = 0.12\ kcf \times 15\ ft\ + 1\ ksf$
$\qquad = 2.8\ ksf$

$u_{sand} = 62.4\ pcf \times 10\ ft$
$\qquad = 624\ psf\ (0.62 ksf)$

$\sigma'_{sand} = \sigma_{sand} - u_{sand}$
$\qquad = 2.8 - 0.62 = 2.2\ ksf$

When it comes to the pore pressure of the clay layer, clay will not drain the excess (now pressurized) water right away (unlike sand). Drainage in this case shall occur over a long period of time instead. The pore pressure of the clay layer will increase due to this by the amount of the added load, and this keeps the effective stress unchanged.

$$\sigma_{clay} = 0.12\ kcf \times 25\ ft + 1\ ksf$$
$$= 4.0\ ksf$$

$$u_{clay} = 0.0624\ kcf \times 20\ ft + 1\ ksf$$
$$= 2.25\ ksf$$

$$\sigma'_{clay} = \sigma_{clay} - u_{clay} = 4.0 - 2.25$$
$$= 1.8\ ksf$$

<u>Over a long period of time after placing the embankment:</u>
The sand effective stress will not change as the water has already drained from it long ago.

$$\sigma'_{sand} = \sigma_{sand} - u_{sand}$$
$$= 2.8 - 0.62$$
$$= 2.2\ ksf$$

As for the clay layer, and over a long period of time, the excess (pressurized) water would have been drained then and this should bring the pore pressure down to:

$$u_{clay} = 0.0624\ kcf \times 20\ ft$$
$$= 1.25\ ksf$$

$$\sigma'_{clay} = \sigma_{clay} - u_{clay}$$
$$= 4.0 - 1.25$$
$$= 2.8\ ksf$$

This makes Profile A the most representative profile.

Correct Answer is (A)

SOLUTION 2.4

The *NCEES Handbook*, Chapter 3 Geotechnical, Section 3.9 Laboratory and Field Compaction is referred to.

Optimum moisture content occurs at the soil's maximum dry density γ_d. In which case, dry density is calculated using the total (moist) density γ_t and moisture content w as follows:

$$\gamma_d = \frac{\gamma_t}{(1 + w)}$$

Density is derived from the volume of the mold provided in the handbook as $1/30\ ft^3$, based upon which, compaction curve is created as shown below:

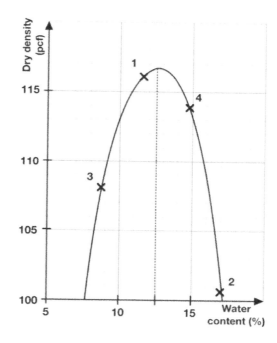

SN	wt_{moist}	γ_t	w	γ_d
	lb	lb/ft³	%	lb/ft³
1	4.32	129.60	11.6%	116.13
2	3.92	117.60	17.0%	100.51
3	3.92	117.60	8.8%	108.09
4	4.36	130.80	14.8%	113.94

The optimum moisture content as derived from the curve is when dry density is max at 12.5%.

Correct Answer is (C)

SOLUTION 2.5

The *NCEES Handbook,* Chapter 3 Geotechnical, Section 3.6 Slope Stability is referred to.

The handbook provides two charts for Taylor (1948). The second is only applicable when friction angle $\emptyset = 0$ and a rock layer has been identified below the slope where the depth factor $D > 1$, which is not the case here.

The first chart – copied in the following page for ease of reference (used with permission from FHWA) – is used in this case.

There are two factors that we shall define prior to performing the calculation:

c_d is the developed, or mobilized, cohesion, which is the cohesion that develops at the slip surface upon failure.

ϕ_d is the developed, or mobilized friction angle, which is the friction angle that develops at the slip surface upon failure.

The safety factor requested in this question represents safety against forming a slip surface, in which case:

$$F.S. = F_c = \frac{c}{c_d}$$

Similarly, the safety factor for the friction angle shall be calculated and shall equal to:

$$F.S. = F_\phi = \frac{\tan \phi}{\tan \phi_d}$$

The above process is iterative in nature, and we may have to perform two or more iterations until the following equation is satisfied:

$$F.S. = F_\phi = F_c$$

Iteration 1: assume $\phi = \phi_d = 20°$

In reference to Taylor's (1948) first chart, using a slope angle $\beta = 55°$ – first iteration shown on the chart – the stability number $N_s = 0.085$.

$$N_s = \frac{c_d}{\gamma H}$$

$\rightarrow c_d = \gamma H N_s$

$= 95 \frac{lb}{ft^3} \times 8\ ft \times 0.085$

$= 64.6\ psf$

$F_c = \frac{c}{c_d} = \frac{58\ psf}{64.6\ psf} = 0.9 = F_\phi$

$$\phi_d = arctan\left(\frac{tan\ \phi}{F_\phi}\right)$$

$$= arctan\left(\frac{tan\ 20°}{0.9}\right)$$

$$= 22°$$

Iteration 2: assume $\phi_d = 22°$

In reference to Taylor (1948) chart, using interpolation → $N_s = 0.077$.

$$c_d = \gamma H N_s$$

$$= 95\ \frac{lb}{ft^3} \times 8\ ft \times 0.077$$

$$= 58.5\ psf$$

$$F_c = \frac{c}{c_d} = \frac{58\ psf}{58.5\ psf} = 0.99 = F_\phi$$

$$\phi_d = arctan\left(\frac{tan\ \phi}{F_\phi}\right)$$

$$= arctan\left(\frac{tan\ 20°}{0.99}\right)$$

$$= 20°$$

It is obvious at this stage that the safety factor falls somewhere between '0.9' to '1.0'.

Correct Answer is (A)

SOLUTION 2.6

The *NCEES Handbook*, Chapter 3 Geotechnical, Section 3.2.1 Normally Consolidated Soils is referred to.

Settlement in a clay layer is calculated using the following equation:

$$S_C = \sum_{1}^{n} \frac{C_c}{1+e_o} H_o\ Log\left(\frac{p_f}{p_o}\right)$$

Where C_c is the compression index and is calculated using the slope of $'Log\ (p) - e'$ graph shown below (*). e_o is initial void ratio

and can be picked up from the graph by substituting for $Log\ (p_o)$. H_o is the layer thickness, p_f is the final pressure and p_o is the initial/original pressure. n in the equation represents the number of layers, in which case the question did not specify more than *one* layer.

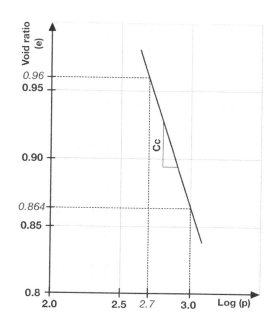

$$C_c = \frac{\Delta e}{\Delta log(p)}\quad (*)$$

$$= \frac{0.96 - 0.864}{3.0 - 2.7}$$

$$= 0.32$$

$$e_o = e_{@\ [log(500)\ =\ 2.7]} = 0.96$$

$$S_C = \frac{C_c}{1+e_o} H_o\ Log\left(\frac{p_f}{p_o}\right)$$

$$= \frac{0.32}{1+0.96} \times 4\ ft \times Log\left(\frac{1,000}{500}\right)$$

$$= 0.2\ ft\ (2.4\ in)$$

Correct Answer is (A)

(*) For the removal of doubt, C_c is an absolute value and is calculated using the delta of void ratio Δe at the numerator.

SOLUTION 2.7

The *NCEES Handbook,* Chapter 4 Geotechnical, Section 3.4.2.1 Bearing Capacity for Concentrically Loaded Square or Rectangular Footings, is referred to in order to provide a solution for this question.

First start with calculating the ultimate bearing capacity for the conditions provided in the question, then have it divided by the acting pressure from the column loading in order to determine the safety factor.

$$q_{ult} = c(N_c)s_c + q(N_q)s_q + 0.5\gamma(B_f)(N_\gamma)s_\gamma$$

The bearing capacity factors N_c, N_q and N_γ are collected from the table provided in the *NCEES Handbook* as 46.1, 33.3 and 48 respectively.

The shape correction factors s_c, s_q and s_γ are calculated using the following equations when $\emptyset > 0$:

$$s_c = 1 + \left(\frac{B_f}{L_f}\right)\left(\frac{N_q}{N_c}\right)$$
$$= 1 + \left(\frac{6}{6}\right)\left(\frac{33.33}{46.1}\right)$$
$$= 1.72$$

$$s_q = 1 + \left(\frac{B_f}{L_f} \tan\emptyset\right)$$
$$= 1 + \left(\frac{6}{6} \tan 35\right)$$
$$= 1.7$$

$$s_\gamma = 1 - 0.4\left(\frac{B_f}{L_f}\right)$$
$$= 1 - 0.4\left(\frac{6}{6}\right)$$
$$= 0.6$$

Given there is no surcharge load:
$$q = \gamma D_f$$
$$= 130 \, pcf \times 5 \, ft$$
$$= 650 \, psf$$

It is also important to remember that the density γ in the bearing capacity equation represents soil at the bottom of the footing, in which case the buoyant one will be used given that the bottom of the footing is submerged. It can either be calculated using the effective stress method as γ' by deducting pore/water pressure from it, or a correction factor of 0.5 can be applied to it which can be collected from the Correction Factor table presented in the same chapter.

$$\gamma' = 130 \, pcf - 62.4 \, pcf$$
$$= 67.6 \, pcf$$

$$q_{ult} = c(N_c)s_c + q(N_q)s_q + 0.5\gamma'(B_f)(N_\gamma)s_\gamma$$
$$= 450(46.1) \times 1.72 + 650(33.3) \times 1.7$$
$$\quad + 0.5 \times 67.6(6)(48) \times 0.6$$
$$= 78,318.54 \, psf$$

$$q_{actual} = \frac{650,000 \, lb}{6 \, ft \times 6 \, ft}$$
$$= 18,055.56 \, psf$$

$$S.F. = \frac{q_{ult}}{q_{actual}}$$
$$= \frac{78,318.54 \, psf}{18,055.56 \, psf}$$
$$= 4.34$$

Correct Answer is (A)

(✼) SOLUTION 2.8

Referring to the *NCEES Handbook version 2.0,* Section 3.8.3:

Dry Density:

$$\gamma_d = \frac{\frac{W_t}{1+w}}{V}$$
$$= \frac{100 \, lb/(1+0.2)}{1 \, ft^3}$$
$$= \mathbf{83.33 \, lb/ft^3}$$

Degree of saturation:

$$S = \frac{w}{\left(\frac{\gamma_w}{\gamma_d} - \frac{1}{G}\right)}$$

$$= \frac{0.2}{\left(\frac{62.4}{83.33} - \frac{1}{2.7}\right)}$$

$$= \mathbf{0.53}$$

Porosity:

$$n = 1 - \frac{W_s}{GV\gamma_w} \left(= 1 - \frac{\gamma_d}{G\,\gamma_w}\right)$$

$$= 1 - \frac{83.33}{2.7 \times 62.4}$$

$$= \mathbf{0.51}$$

SOLUTION 2.9
The *NCEES Handbook,* Chapter 3 Geotechnical, Section 3.7.4 Rock Classification is referred to.

Rock Quality Designation RQD is measured as follows:

$$RQD = \frac{\Sigma\, Length\ of\ Sound\ Core\ Pieces > 4\ in}{Total\ Core\ Run\ Legnth}$$

$$= \frac{70\ ft}{100\ ft}$$

$$= 70\%$$

Per the description provided in the FHA Soils and Foundation reference manual, which can be found in the *NCEES Handbook,* an *RQD* of 70% has a fair quality.

Correct Answer is (C)

SOLUTION 2.10
Reference is made to the *NCEES Handbook version 2.0,* Section 3.1.2.

At rest Rankine Coefficient for normally consolidated soils:

$$K_{o,NC} = 1 - sin\emptyset'$$

$$0.33 = 1 - sin\,\emptyset' \quad \rightarrow sin\,\emptyset' = 0.67$$

For over consolidated soils:
$$K_{o,OC} = (1- sin\emptyset') \times OCR^{sin\emptyset'}$$

$$= K_{o,NC} \times OCR^{sin\emptyset'}$$

$$OCR = \left(\frac{K_{o,OC}}{K_{o,NC}}\right)^{\frac{1}{sin\emptyset'}}$$

$$= \left(\frac{0.85}{0.33}\right)^{\frac{1}{0.67}}$$

$$= 4.1$$

Correct Answer is (A)

SOLUTION 2.11
Generally, well graded Sands have higher friction angles compared to gap graded sands.

From the presented chart, Sand 2 seems to have gaps in its gradation around particle sizes 0.1 *mm* to 1.0 *mm,* and this will have a detrimental effect on its friction angle.

Shear strength is proportional to cohesion and to the friction angle, see below:

$$\tau = c + \sigma_n tan\emptyset$$

τ Shear strength

c Total cohesion

σ_n Normal stress

$\emptyset$ Friction angle

It is therefore more likely that sand 1 will have a higher friction angle compared to Sand 2, which renders Sand 1 stronger in shear during normal loading.

Correct Answer is (B)

SOLUTION 2.12

Clay is an undrained layer, which means that when loaded, water will not drain immediately. Rather, water, due to excess pressure, will drain/get expelled slowly and over a long period of time. The slow expulsion of water from the voids between clay particles causes the layer to lose its structure which leads into its ultimate and slow settlement over time.

When clay is loaded, and due its undrained property, pore water pressure increases. Upon water expulsion, pore water pressure decreases over time resulting in a gradual long-term increase in effective stress.

Correct Answer is (B)

SOLUTION 2.13

NCEES Handbook version 2.0 Section 3.7 provides a guideline for the minimum number of exploration points and their depth, as published by the *Federal Highway Administration* FHWA 2002.

The guideline stipulates that the minimum number of boreholes required for retaining walls $> 100\ ft$ in length should be spaced $100\ ft$ to $200\ ft$ with locations alternating from the front of the wall to behind it. The depth of these exploration points should be 1 to 2 times the wall height or a minimum of $10\ ft$ below the bedrock.

Correct Answer is (B)

📖 SOLUTION 2.14

The *shrinkage limit* represents the water content that corresponds to transitioning between a brittle state and a semi-solid state.

The *plastic limit* represents the water content that corresponds to transitioning between a semi-solid state and a plastic state.

The *liquid limit* represents the water content that corresponds to transitioning from a plastic state to a liquid state.

The *plasticity index* represents the range within which soil remains in its plastic state bounded with its *plasticity limit* as the lower limit, and the *liquid limit* as the upper limit to this range.

This renders the *liquid limit* the maximum water content any soil can get to.

Correct Answer is (A)

SOLUTION 2.15

Using the AASHTO classification system found in the *NCEES Handbook version 2.0* Section 3.7.3, nearly 45% is finer than $0.075\ mm$ (i.e., passing No. 200 sieve). This indicates that the sample is not granular and falls within the Silt-Clay material categories of: A-4, 5, 6 or 7.

The characteristics of fines – material finer than $0.425\ mm$ (i.e., passing No. 40 sieve) were given as follows:

$$Liquid Limit (LL) = 50$$
$$Plastic limit (PL) = 35$$
$$Plasticity\ Index\ (PI) = 50 - 35 = 15$$

Based on this, the material is classified as either A-7-5 or A-7-6.

Using the comment section of the classification table and provided that $LL - 30 = 20 > PI$, classification of the material would be that of A-7-5.

The group index GI for this category is calculated as follows:

GI = (F − 35) [0.2 + 0.005 (LL − 40)]
    + 0.01 × (F − 15)(PI − 10)

= (45 − 35) [0.2 + 0.005 (50-40)]
    + 0.01 × (45 − 15)(15 − 10)

= 4

Final classification is A-7-5 (4)

Correct Answer is (A)

SOLUTION 2.16
The following information is gathered from the gradation chart:

Nearly 45% is finer than 0.075 mm (No. 200 sieve), which means 55% is retained on this Sieve classifying the sample as either sand or gravel.

Nearly 82% is finer than 4.75 mm (No. 4 sieve), which means that 18% is retained on this sieve classifying the sample as sand.

Based on the USCS classification system found in the NCEES Handbook version 2.0 Table 3.7.2, Soils with the following:

> 50% retained on sieve No. 200
> 50% passes No. 4 sieve

Those can either be clean sands (either SW or SP), or sands with fines (either SM or SC).

Since fines are > 12% ($i.e.$, 45%), this can be classified as SM or SC.

Since the question indicated the fines are classified as MH, this sand can be classified as SM.

Correct Answer is (D)

SOLUTION 2.17
The weight of cement consumed to produce 1 ft^3 is calculated. Based upon which, the quantity of water can be determined based on the w/c ratio provided.

One bag of cement is 94 lb and is sufficient to produce the following yield:

Component	Ratio	Weight	Volume
		lb	ft^3
Cement	1	94	0.48
Fine agg.	1.5	141	0.85
Coarse agg.	3	282	1.70
Water		42.3	0.68
TOTAL			3.71

Volume of water required for 1 ft^3
= 0.68/3.71 = 0.183 ft^3 (5.18 $litres$)

Add an extra 5% to account for aggregates:

= 5.4 $litres$

Correct Answer is (A)

SOLUTION 2.18
The flexible wall permeameter test is used when the tested materials' permeability is lower than $1 \times 10^{-3} cm/sec$. The specimen in this case is encased in a membrane, and with the proper amount of pressure, flow through the specimen is recorded with time.

Correct Answer is (D)

SOLUTION 2.19
Sulphate Resisting Cement Type V is the preferred option when used in sewage treatment plants because of its high resistance to sulphate attacks that is generated from sewage.

Correct Answer is (D)

SOLUTION 2.20
A test pressure that is no less than 150 psi or 1.25 times the operating pressure shall be applied for 2 hours.

Correct Answer is (D)

SOLUTION 2.21

One of the important factors in deciding pipeline material is its installation.

Embedment (i.e., pipeline bedding) is an important platform for flexible pipelines such as HDPE, steel, or fiberglass pipes. Because of this, continuous or fulltime inspection for bedding material shall be provided.

When supervision is not available, a rigid pipe material option shall be selected for the project, in which case, **concrete pipes** are most suitable.

Correct Answer is (A)

SOLUTION 2.22

The *Recommended Standards for Water Works Facilities, 2018,* Section 4.4.6.6 Pipe Materials (for disinfection by ozone) recommends the use of low carbon $304L$ and $316L$ with the latter being the preferred option.

Correct Answer is (B)

III
HYDRAULICS
CLOSED CONDUIT & OPEN CHANNEL

Knowledge Areas Covered

SN	Knowledge Area
5	**Hydraulics – Closed Conduit** A. Energy and/or continuity equation (e.g., Bernoulli, grade line analyses, momentum equation) B. Pressure conduit (e.g., single pipe, force mains, Hazen-Williams, Darcy-Weisbach, major and minor losses) C. Pump application and analysis, including wet wells, lift stations, and cavitation. D. Pipe network analysis (e.g., series, parallel, loop networks)
6	**Hydraulics – Open Channel** A. Open channel flow B. Hydraulic grade lines and energy dissipation (e.g., plunge pool, drop structure, culvert outlet) C. Stormwater collection and drainage (e.g., culvert, stormwater inlets, gutter flow, street flow, storm sewer pipes) D. Sub- and supercritical flow

PART III
Hydraulics

PROBLEM 3.1 *Water Surface Elevation*

The below elevated water reservoir is being drained into the atmosphere as shown using a 12 *in* pipe. The flow rate of water has been measured as 7,500 *gpm* (gallon per minute)

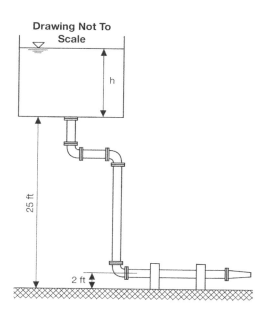

Assuming head losses through this system equal to 25 *ft*. The depth of the water in the reservoir h is most nearly:

(A) 3 *ft*

(B) 5 *ft*

(C) 7 *ft*

(D) 9 *ft*

PROBLEM 3.2 *Head Losses*

The head loss in *ft* for a 20-year-old cast iron pipe, 200 *ft* long, 20 *in* dia, with a slope of 2%, is most nearly:

(A) 9

(B) 4

(C) 20

(D) 2

PROBLEM 3.3 *Elevation of Water Surface in Two Reservoirs*

A galvanized iron (*) pipeline with two 45° bends connects two large reservoirs as shown below. The diameter of the pipe is 12 *in* and is discharging 80° F water at a rate of 20 *cfs*.

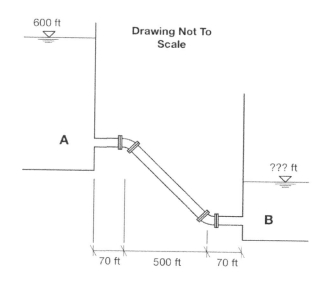

Given the elevations and distances in the above figure, elevation of the water surface at reservoir 'B' is most nearly:

(A) 414 *ft*

(B) 190 *ft*

(C) 553 *ft*

(D) 50 *ft*

(*) Use upper value of roughness (ε) for galvanized iron pipes.

PROBLEM 3.4 *Vertical Force at a Joint of a Jet*

Jet nozzles at 'B' and 'C' with equal diameters of 2 in discharge water at a flow rate of 0.5 cfs per nozzle into the open.

The supply pipe 'A' has the same diameter and water is forced into it at a pressure of 75 psi.

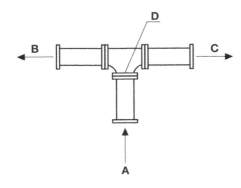

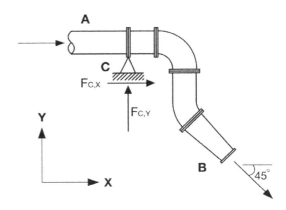

Ignoring the weight of water and the pipe arrangement, the vertical force that holds joint 'D' into place is most nearly:

(A) 90 lb

(B) 150 lb

(C) 235 lb

(D) 325 lb

(⁂) PROBLEM 3.5 *Water Discharge External Forces*

An inclined nozzle at 'B' with a diameter of 1.5 in, discharges 1 cfs water into the atmosphere.

The supply pipe 'A' has a diameter of 3.5 in, and the nozzle is held into place by hinge 'C' as shown below.

Ignoring the weight of water and the pipe arrangement, the horizontal and vertical reactions at support 'C' when pressure at 'A' is 100 psi are:

(A) $F_{C,x} = 879.4\ lb \leftarrow$
$F_{C,y} = 111.8\ lb \downarrow$

(B) $F_{C,x} = 77.25\ lb \rightarrow$
$F_{C,y} = 113.0\ lb \downarrow$

(C) $F_{C,x} = 837.0\ lb \leftarrow$
$F_{C,y} = 176.7\ lb \downarrow$

(D) $F_{C,x} = 132.0\ lb \leftarrow$
$F_{C,y} = 22.5\ lb \downarrow$

PROBLEM 3.6 *Gate under Pressure*

The below is an inclined trapezoidal gate fixed with a hinge at its top as shown and is blocking a pathway for a water course.

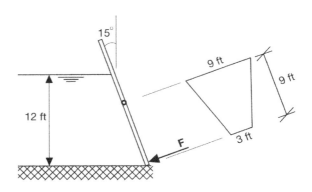

Ignoring the weight of this gate, the required force F that can hold this gate in place is most nearly:

(A) 3 kip

(B) 6 kip

(C) 12 kip

(D) 23 kip

PROBLEM 3.7 *Hazen-Williams Coefficient Determination*

The equivalent *Hazen-Williams* Coefficient for a 12 in diameter 1 mile long pipeline with a drop of 110 ft in energy line, flowing full and with a *Manning* coefficient of $n = 0.013$, is most nearly:

(A) 86

(B) 96

(C) 106

(D) 116

PROBLEM 3.8 *Increase in Pipe Flow due to Diameter Increase*

An increase in pipe diameter from 0.5 in to 2 in while maintaining the same pressure gradient would increase the laminar flow rate by a factor of:

(A) 4

(B) 16

(C) 64

(D) 256

PROBLEM 3.9 *Water Hammer Pressure*

Using a wave speed of 3,200 ft/sec, a 3,000 ft pipe 24 in diameter with a flow rate of 15 cfs experiences a change in pressure of _____ lb/ft² when its control valve closes in 4 seconds.

PROBLEM 3.10 *Head Increase Required by a Pump*

The below pump generates a flow rate of 10 cfs in the shown closed horizontal two loop network. Each loop is 100 ft by 100 ft of length. All pipes are 12 in diameter, and the lower range for roughness (ε) for cast iron pipes can be used for friction coefficient calculation.

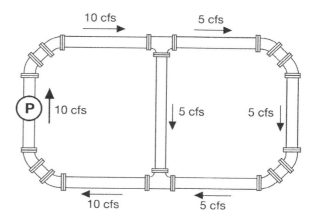

The head increase that the pump is required to provide to keep this flow constant at a temperature of 80° F, assuming the branch flow rates are accurate, is most nearly:

(A) 30 ft

(B) 8 ft

(C) 15 ft

(D) 45 ft

PROBLEM 3.11 *Packed Bed Design*

A 2 ft deep packed bed reactor with spherical particles of 0.01 ft diameter, shape factor of 1.0, and porosity of 0.45. The reactor receives 25° C water flowing at a rate of 5 cfs. If the pressure loss is to be limited to 40 kPa, the bed cross-section should be:

(A) 1.5 ft²

(B) 2.5 ft²

(C) 3.5 ft²

(D) 4.5 ft²

PROBLEM 3.12 *Pump Required Power*

In the below elevated tank, pump P pumps water at a rate of 2 MGD through a 36 in diameter steel pipe.

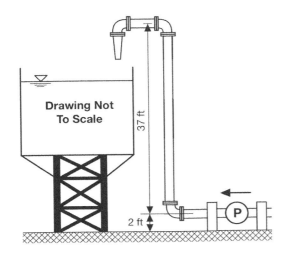

Assuming the total head losses through this pipe equal to 12.5 ft, and a pump efficiency of 75%, the required pump horsepower is most nearly:

(A) 23 hp

(B) 8 hp

(C) 15 hp

(D) 45 hp

PROBLEM 3.13 *Pump Upgrade*

An existing pump that pushes a flow rate of 0.5 MGD at a rotational speed of 1,750 rpm with an impeller size of 10 in is upgraded into a rotational speed of 2,250 rpm and an impeller size of 12 in. The new flow rate is expected to be:

(A) 0.6 MGD

(B) 0.7 MGD

(C) 0.9 MGD

(D) 1.1 MGD

PROBLEM 3.14 *Pump Elevation*

Pump P below pumps 80° F water at a rate of 10 cfs through a 12 in pipe with an elevation difference of 50 ft as shown below.

The cavitation parameter for the pump is 0.15 and the pressure on top of the two reservoirs is atmospheric pressure (*).

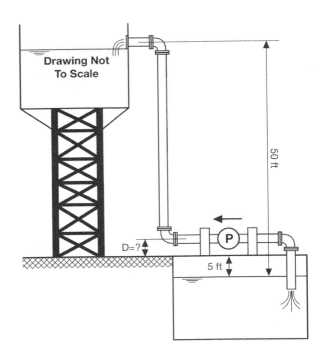

Assuming total head losses at the suction side of 4 ft and the rest of the pipe inclusive of all bends and exits equals to 27.5 ft, the maximum allowable distance the pump shall be placed on top of the ground surface (D) is most nearly:

(A) 9 ft

(B) 14 ft

(C) 34 ft

(D) 4 ft

(*) use $P_{atm} = 14.696\ psi$

PROBLEM 3.15 *Loop Network Flow Calculation*

The below water supply network consists of 12 *in* diameter steel pipes with a friction factor of $f = 0.02$ for all pipes.

The incoming flow to point A is $10\,cfs$, however branch flows of $6\,cfs$ and 4 cfs shown on the below network are only estimates.

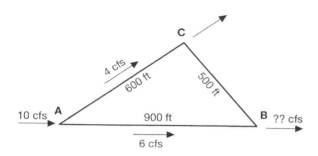

Using the *Hardy-Cross* algorithm, the correction factor has been determined as $\Delta Q = +1.44$ cfs assuming positive for clockwise flows. Based on this, the outgoing flow from point B is most nearly:

(A) The correction factor is incorrect.

(B) $6\,cfs$

(C) $4.54\,cfs$

(D) $7.44\,cfs$

PROBLEM 3.16 *Parallel Network Flow Calculation*

The below water supply network consists of 12 *in* diameter for pipeline ACB and a friction factor of $f = 0.02$ for all pipes.

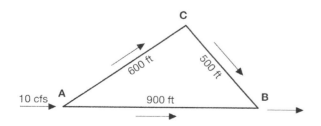

The diameter for pipe AB that makes the flow rate in pipe ACB five times the flow rate of pipe AB is:

(A) $4\,in$

(B) $6\,in$

(C) $8\,in$

(D) $10\,in$

PROBLEM 3.17 *Seawater Canal*

The below seawater return canal is made of concrete lining and is situated in an industrial area. Its purpose is to return seawater used to cool down equipment from factories in this industrial area back to the sea through an outfall.

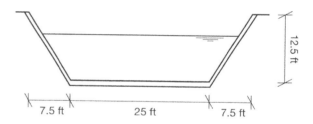

Assuming a continuous daily operation of this canal with a steady uniform flow, a design freeboard of $3.75\,ft$ and a longitudinal slope of 2%, the seawater intake structure should be sized for a maximum intake (in MGD) of nearly:

(A) 9,000

(B) 7,750

(C) 5,300

(D) 9,200

PROBLEM 3.18 Water Channel

The below is a cross section for a V shaped open channel that delivers water at 70° F with a 1 ft free board. The mean velocity of the flow is 0.3 ft/sec.

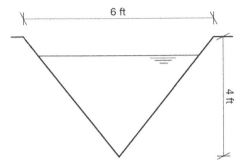

Reynolds number for this flow is most nearly:

(A) 0.32

(B) 32,000

(C) 0.26

(D) 26,000

PROBLEM 3.19 Flow in a Gutter

The below is a gutter with vertical curb located to the side of a road and is made of concrete (*).

The longitudinal slope of this gutter is 0.01 ft/ft.

The maximum flow this gutter can take during a storm event without the water overtopping the curb is most nearly:

(A) 6 cfs

(B) 36 cfs

(C) 1 cfs

(D) 235 cfs

(*) Use a *Manning* coefficient for concrete of '0.013'.

PROBLEM 3.20 Culvert Inlet Headwater Elevation

A 100 ft long, 33 in diameter, concrete pipe culvert with a square/flushed cut end, having its inlet invert elevation at 317 ft, outlet invert elevation of 313 ft, has its <u>inlet</u> control headwater elevation at _____ ft when subjected to a 40 cfs flow:

(A) 363 ft

(B) 321 ft

(C) 320 ft

(D) 325 ft

PROBLEM 3.21 Culvert Outlet Headwater Elevation

A 200 ft long, 33 in diameter, concrete pipe culvert with a square/flushed cut end, having its inlet invert elevation at 317 ft, outlet invert elevation of 313 ft and a tailwater depth at outlet of 3.5 ft, has its <u>outlet</u> control headwater elevation at _____ ft when subjected to a 40 cfs flow:

(A) 318.6 ft

(B) 320.1 ft

(C) 322.6 ft

(D) 325.1 ft

PROBLEM 3.22 Hydraulic Jump

A hydraulic jump is occurring in an 8 ft wide rectangular channel with a discharge rate of 80 cfs and an initial water depth of 1 ft before the jump.

The depth of the water right after the hydraulic jump is most nearly:

(A) 3.0 ft

(B) 2.5 ft

(C) 2.0 ft

(D) 1.5 ft

PROBLEM 3.23 *Energy Loss*
The energy loss which occurs due to a hydraulic jump in a water channel from an initial depth of 1 ft to a final depth right after the jump of 4 ft is most nearly:

(A) 0.6 ft

(B) 1.7 ft

(C) 2.7 ft

(D) 3.0 ft

PROBLEM 3.24 *Half Circular Channel*
The below is an open channel made of a half concrete 72 in diameter pipe along with a straight concrete edge built on top of it as an extra 3 ft on each side.

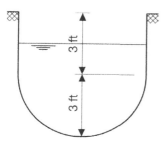

Considering a free board of 1.5 ft only, a *Chezy* resistance coefficient of '55' and a slope of 0.01 ft/ft, the expected flow in the channel in MGD is most nearly:

(A) 173

(B) 112

(C) 125

(D) 80

PROBLEM 3.25 *Pipe Flow Depth*
A flow rate of 10 cfs of storm drain is conveyed using a concrete pipe at a slope of 0.005 ft/ft. The *Manning* coefficient for the concrete pipe is 0.015.

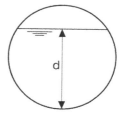

Using a pipe diameter of 30 in, the depth d of the liquid inside this pipe is most nearly:

(A) 30 in

(B) 20 in

(C) 10 in

(D) 15 in

PROBLEM 3.26 *Culvert Flow*
The below is a 4 ft diameter concrete 75 ft long culvert with a coefficient of discharge of 0.95 and a *Manning* coefficient of 0.015.

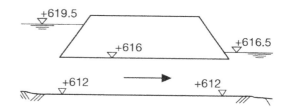

The total discharge through this culvert in ft^3/sec is most nearly:

(A) 138 cfs

(B) 165 cfs

(C) 262 cfs

(D) 114 cfs

PROBLEM 3.27 *Critical Slope*
A rectangular channel 5 ft wide and 4 ft deep carries a flow of 65 cfs and has a *Manning* coefficient of $n = 0.013$ has the following critical slope:

(A) 0.4%

(B) 2.3%

(C) 2.9%

(D) 3.3%

(A) 0.5 ft

(B) 0.9 ft

(C) 1.5 ft

(D) 2.1 ft

PROBLEM 3.28 *Hydraulic Radius Composite Channel*

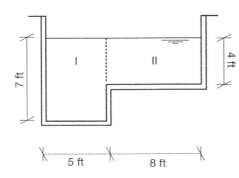

The hydraulic radius for Part I of the above open water channel is most nearly:

(A) 1.8 ft

(B) 2.3 ft

(C) 2.5 ft

(D) 1.4 ft

PROBLEM 3.29 *Depth of Flow*

The below water canal is made out of concrete with a *Manning* coefficient of $n = 0.013$ and a slope of $0.001\ ft/ft$

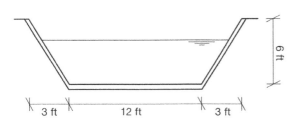

The expected depth of flow when the flow rate is 80 cfs is most nearly:

📖 PROBLEM 3.30 *Gradually Varied Flow*

The below are longitudinal sections for a wide rectangular channel with a sluice gate that has a 3 ft vertical opening. The channel starts with a slope of $0.02\ ft/ft$ and continues with a slope of $0.001\ ft/ft$ as shown in the below options (slope exaggerated for demonstration purposes only).

The best option that represents a flow rate of 40 cfs per feet of width (*) is:

(A) Option A

(B) Option B

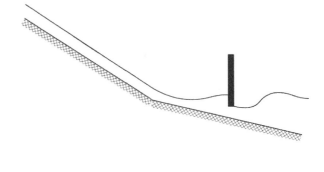

(C) Option C

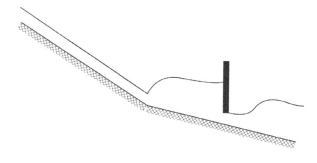

(D) Option D

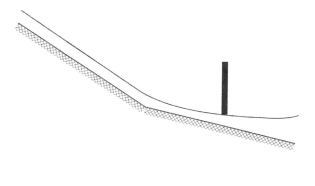

(*) Use the following depth formulas for wide channels and for this channel:

Critical depth

$$y_c = \left(\frac{q^2}{g}\right)^{1/3}$$

Normal depth

$$y_n = \left(\frac{0.025q}{S^{1/2}}\right)^{3/5}$$

Where:

q is in $ft^3/\sec$ per feet

S is the longitudinal channel slope in ft/ft

g is 32.174 ft/sec^2

PROBLEM 3.31 *Roughness Coefficient*

The below water canal is made of different materials each has diferent roughness coefficient as dispalyed in the below table:

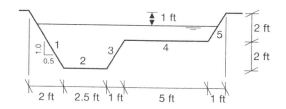

Component	Material used	Coefficient
1	Concrete	0.011
2	Graveled surface	0.03
3	Bare sand	0.01
4	Grass	0.45
5	Bare sand	0.01

The composite roughness coefficient is:

(A) 0.11

(B) 0.36

(C) 0.17

(D) 0.28

PROBLEM 3.32 *Culvert Flow Velocity*

The below 100 ft long 6 ft diameter culvert has an entrance loss coefecient of 0.5, $n = 0.015$, and an upstream velocity of 15 ft/sec.

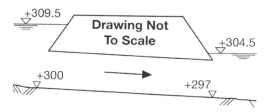

Given the above information, downstream flow velocity is most nearly:

(A) 8 ft/sec

(B) 11 ft/sec

(C) 13 ft/sec

(D) 17 ft/sec

SOLUTION 3.1

The energy equation found in the *NCEES Handbook* Section 6.2.1.2 is used to solve this problem.

If the elevation at the reservoir surface is A, and at the pipe outlet location is B, the following can be established:

$$\frac{P_A}{\gamma} + z_A + \frac{v_A^2}{2g} = \frac{P_B}{\gamma} + z_B + \frac{v_B^2}{2g} + h_f$$

Where:

$$P_A = 0$$
$$P_B = 0$$
$$v_A = 0$$

$$v_B = \frac{Q}{A} = \frac{7{,}500 \frac{gal}{min} \times \frac{0.134 \frac{ft^3}{gal}}{60 \frac{sec}{min}}}{\pi \times (0.5)^2 \, ft^2} = 21.3 \, ft/sec$$

Also, when taking datum at ground level:

$$z_A = 25 + h$$
$$z_B = 2 \, ft$$

Substitute the above into the energy equation as follows:

$$\frac{P_A}{\gamma} + z_A + \frac{v_A^2}{2g} = \frac{P_B}{\gamma} + z_B + \frac{v_B^2}{2g} + h_f$$

$$0 + (25 + h) + 0 = 0 + 2 + \frac{(21.3)^2}{2 \times 32.174} + 25$$

$$\rightarrow h = 9 \, ft$$

Correct Answer is (D)

SOLUTION 3.2

Using the *Hazen-Williams* equation and coefficients provided in Section 6.3 of the *NCEES Handbook version 2.0*, see below:

$$h_f = \frac{4.73 \, L}{C^{1.852} \, D^{4.87}} \, Q^{1.852}$$

C is the *Hazen-Williams* Coefficient which equals to 100 for 20-year-old pipes (See Section 6.3.1.4). L and D are length and diameter in ft.

$$Q = 1.318 \, C A R_H^{0.63} \, S^{0.54}$$

R_H is the hydraulic radius, and it equals to the area of flow (A) divided by the wetted perimeter (P). For pipelines running in full capacity, $R_H = r/2 = D/4$

$$Q = 1.318 \times 100 \times \pi \times \left(\frac{10}{12} ft\right)^2 \times \left(\frac{5}{12} ft\right)^{0.63} \times 0.02^{0.54}$$

$$= 20 \, ft^3/sec$$

$$h_f = \frac{4.73 \times 200}{100^{1.852} \times \left(\frac{20}{12}\right)^{4.87}} \times 20^{1.852}$$

$$= 4.0 \, ft$$

A quicker way for determining the answer is by using the slope S as follows:

$$S = h_f / L$$

$$\rightarrow h_f = S \times L = 0.02 \times 200 = 4 \, ft$$

Correct Answer is (B)

SOLUTION 3.3

The *NCEES Handbook,* Section 6.3 Closed Conduit Flow and Pumps can be used to solve this problem.

Given the inputs in this question, the following equations/sections are referred to:

- The Energy Equation – Section 6.2.1.2
- *Reynolds* number for circular pipes – Section 6.2.2.2
- *Darcy-Weisbach* equation for head losses due to flow – Section 6.2.3.1
- Minor Losses in Pipe Fittings, Contractions and Expansions – Section 6.3.3

The Energy Equation:

$$\frac{P_A}{\gamma} + z_A + \frac{v_A^2}{2g} = \frac{P_B}{\gamma} + z_B + \frac{v_B^2}{2g} + h_f$$

Pressure at surface A and surface B is zero (atmospheric only). Also, velocity of the surface dropping in level can be neglected in large reservoirs.

With this, the energy equation can be reduced to:

$$z_A = z_B + h_f$$

Calculating Head Losses h_f:

Head loss due to flow:

Using *Darcy-Weisbach* equation for head losses:

$$h_{f,flow} = f \frac{L}{D} \frac{v^2}{2g}$$

f is a function of R_e and the relative roughness $\left(\frac{\varepsilon}{D}\right)$ taken from the *Moody, Darcy, or Stanton* Friction Factor Diagram page 312 of the *NCEES Handbook version 2.0*.

v in the denominator of *Reynolds* number equation (shown below) is the kinematic viscosity taken from the physical properties of water table Section 6.2.1.6 as $0.93 \times 10^{-5} \, ft^2/sec$.

v in the numerator is velocity in the pipe

$$= \frac{Q}{A} = \frac{20 \, ft^3/sec}{\pi (0.5 \, ft)^2} = 25.5 \, ft/sec$$

$$R_e = \frac{vD}{v}$$

$$= \frac{25.5 \, ft/sec \times 12 \, in \times \frac{1 \, ft}{12 \, in}}{0.93 \times 10^{-5} \, ft^2/sec}$$

$$= 2.7 \times 10^6$$

$$\frac{\varepsilon}{D} = \frac{0.0008}{1 \, ft} = 0.0008$$

Substituting R_e and $\frac{\varepsilon}{D}$ in *Moody, Darcy, or Stanton* Friction Factor Diagram page 312 of the *NCEES Handbook version 2.0* generates a friction value of $f = 0.019$.

$$h_{f,flow} = f \frac{L}{D} \frac{v^2}{2g}$$

$$= 0.019 \times \frac{\left(70 + \frac{500}{\sin(45)} + 70\right) ft}{1 \, ft} \frac{\left(25.5 \frac{ft}{sec}\right)^2}{2 \times 32.174 \frac{ft}{sec^2}}$$

$$= 162.6 \, ft$$

Head loss due to pipe fittings:

$$h_{f,fittings} = C \frac{v^2}{2g}$$

Where C equals to '1.0', '0.5' and '0.4' for sharp reservoir exit, sharp reservoir entrance and 45° elbow respectively taken from Sections 6.3.3.1 and 6.3.3.2.

$$h_{f,fittings} = C_{total} \frac{v^2}{2g}$$

$$= (1 + 2 \times 0.4 + 0.5) \times \frac{\left(25.5 \frac{ft}{sec}\right)^2}{2 \times 32.174 \frac{ft}{sec^2}}$$

$$= 23.2 \, ft$$

Energy Equation and Total Head Losses:

$$z_B = z_A - h_f$$

$$= z_A - (h_{f,flow} + h_{f,fittings})$$

$$= 600 - (162.6 + 23.2)$$
$$= 414.2 \, ft$$

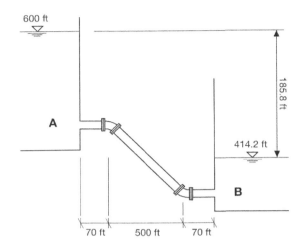

Correct Answer is (A)

SOLUTION 3.4
The impulse momentum principle is used to determine forces in the vertical direction y to compute the force exerted at the cap.

$$Q_A = Q_B + Q_C = 1 \, cfs$$

$$V_A = \frac{Q_A}{A_A} = \frac{1 \, ft^3/sec}{\pi \times (1/12)^2 \, ft^2} = 45.8 \, ft/sec$$

$$\sum F_y = \frac{\gamma \times Q \times \Delta V_{A,y}}{g_c}$$

$$= \frac{\gamma \times Q \times (0 - V_{A,y})}{g_c}$$

$$= \frac{62.4 \, pcf \times 1 \, cfs \times (0 - 45.8 \, ft/sec)}{32.174 \, lbm.ft/lbf.sec^2}$$

$$= -88.8 \, lb$$

$$F_A - F_{cap} = -88.8 \, lb$$

$$P \times A_A - F_{cap} = -88.8 \, lb$$

$$75 \frac{lb}{in^2} \times (\pi \times 1^2) \, in^2 - F_{cap} = -88.8 \, lb$$

$$\rightarrow F_{cap} = 324.4 \, lb \downarrow$$

Correct Answer is (D)

(✻) SOLUTION 3.5
The impulse momentum principle is used to determine forces F_A and F_B which corresponds to the reactions at the support.

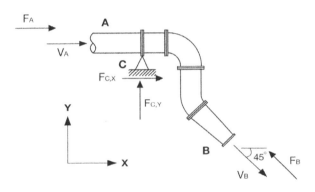

$$V_A = \frac{Q}{A_A} = \frac{1 \, cfs}{\pi \times (1.75/12)^2 \, ft^2} = 15.0 \, ft/sec$$

$$V_B = \frac{Q}{A_B} = \frac{1 \, cfs}{\pi \times (0.75/12)^2 \, ft^2} = 81.5 \, ft/sec$$

$$\sum F_x = \frac{\gamma \times Q \times \Delta V_x}{g_c}$$

$$= \frac{\gamma \times Q \times (V_{B,x} - V_{A,x})}{g_c}$$

$$= \frac{62.4 \, pcf \times 1 \, cfs \times (81.5 \cos(45) - 15)}{32.174 \, lbm.ft/lbf.sec^2}$$

$$= 82.7 \, lb$$

$$F_A - F_B \times cos(45) + F_{C,x} = 82.7 \; lb$$

$$F_{C,x} = 82.7 \; lb - F_A + F_B \times cos(45)$$
$$= 82.7 \; lb - P_A A_A + (P_B A_B) \times cos(45)$$
$$= 82.7 - 100 \times 1.75^2 \pi + 0$$
$$= -879.4 \; lb \; \leftarrow$$

$$\sum F_y = \frac{\gamma \times Q \times \Delta V_y}{g_c}$$
$$= \frac{\gamma \times Q \times (V_{B,y} - V_{A,y})}{g_c}$$
$$= \frac{62.4 \; pcf \times 1 \; cfs \times (-81.5 \; sin(45) - 0)}{32.174 \; lbm.ft/lbf.sec^2}$$
$$= -111.8 \; lb$$

$$F_B \times sin(45) + F_{C,y} = -111.8 \; lb$$

$$F_{C,y} = -111.8 \; lb - F_B \times sin(45)$$
$$= -111.8 \; lb - P_B \times A_B \times sin(45)$$
$$= -111.8 \; lb - 0 \; psi \times 0.75^2 \pi \; sin(45)$$
$$= -111.8 \; lb \downarrow$$

Correct Answer is (A)

SOLUTION 3.6
Reference is made to the *NCEES Handbook* Section 6.1.4.4 Forces on Submerged Surfaces and Center of Pressure.

Start with finding the centroid $\bar{y}$ for the trapezoidal shape along with the moment of inertia about the centroid I_{xC} using the handbook's first chapter:

$$\bar{y} = \frac{9 \times (2 \times 3 + 9)}{3 \times (3 + 9)} = 3.75 \; ft$$

$$I_{xC} = \frac{9^3(3^2 + 4 \times 3 \times 9 + 9^2)}{36 \times (3 + 9)} = 334.1 \; ft^4$$

$$A = \frac{9 \times (3 + 9)}{2} = 54 \; ft^2$$

To proceed with the solution the slant distance y_C to the centroid of the trapezoidal shape along with the slant distance to the center of pressure y_{CP} shall be calculated.

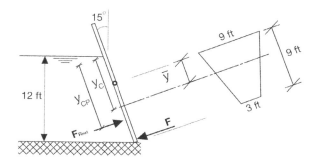

Using similarity of triangles, y_C can be determined as follows:

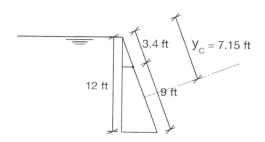

$$y_C = \frac{12}{cos(15)} - 9 + 3.75 = 7.15 \; ft$$

Based on the above, y_{CP} can be calculated using the same section of the handbook as follows:

$$y_{CP} = y_C + \frac{I_{xC}}{y_C A}$$
$$= 7.15 + \frac{334.1}{7.15 \times 54}$$
$$= 8 \; ft$$

The resultant pressure force F_{Rnet} is calculated using the area of the gate and the vertical distance to the centroid of the gate (in which case a cosine to the given angle is used given that the gate in inclined outwards as opposed to inwards as given in the handbook). See below:

$$F_{Rnet} = (\rho g y_c \cos\theta)A$$

$$= 1.94 \frac{lbf.sec^2}{ft^4} \times 32.174 \frac{ft}{sec^2} \times 7.15\,ft \times \cos(15) \times 54\,ft^2$$

$$= 23{,}278.3\,lbf$$

Taking moment around the hinge to determine force F:

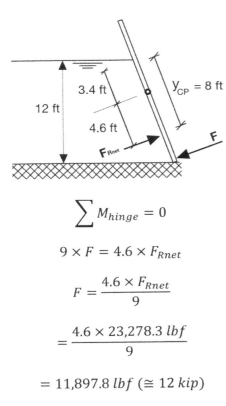

$$\sum M_{hinge} = 0$$

$$9 \times F = 4.6 \times F_{Rnet}$$

$$F = \frac{4.6 \times F_{Rnet}}{9}$$

$$= \frac{4.6 \times 23{,}278.3\,lbf}{9}$$

$$= 11{,}897.8\,lbf\,(\cong 12\,kip)$$

Correct Answer is (C)

SOLUTION 3.7
Section 6.3.1 and 6.4.5.1 of the *NCEES Handbook* are referred to as follows:

Flow using Manning equation:

$$Q = \frac{1.486}{n} AR_H^{2/3} S^{1/2}$$

Flow using Hazen-Williams equation:

$$Q = 1.318 CAR_H^{0.63} S^{0.54}$$

Equating the two equations:

$$1.318 CAR_H^{0.63} S^{0.54} = \frac{1.486}{n} AR_H^{2/3} S^{1/2}$$

$$C = \frac{1.486 AR_H^{2/3} S^{1/2}}{1.318 n AR_H^{0.63} S^{0.54}}$$

$$= \frac{1.127}{n} \times (R_H)^{\frac{2}{3} - 0.63} \times (S)^{\frac{1}{2} - 0.54}$$

$$= \frac{1.127}{n} \times (R_H)^{0.037} \times (S)^{-0.04}$$

$$= \frac{1.127}{0.013} \times \left(\frac{A}{P}\right)^{0.037} \times \left(\frac{h_f}{L}\right)^{-0.04}$$

$$= \frac{1.127}{0.013} \times \left(\frac{r}{2}\right)^{0.037} \times \left(\frac{h_f}{L}\right)^{-0.04}$$

$$= \frac{1.127}{0.013} \times \left(\frac{0.5\,ft}{2}\right)^{0.037} \times \left(\frac{110\,ft}{5{,}280\,ft}\right)^{-0.04}$$

$$= \frac{1.127}{0.013} \times 0.95 \times 1.167$$

$$\cong 96$$

Correct Answer is (B)

SOLUTION 3.8
Given the flow is laminar, the *Hagen-Poiseuille* equation of Section 6.2.3.2 of the *NCEES Handbook version 2.0* is used as follows:

$$Q = \frac{\pi D^4 \Delta P_f}{128 u L}$$

Constant factors such as viscosity u, length L, π, and pressure difference/gradient ΔP_f are removed from the equation for ease of handling as follow:

$$Q \propto D_2^4$$

Given $D_1 = 0.5\ in$ and $D_2 = 2\ in$, then $D_2 = 4 \times D_1$

$$Q \propto (4D_1)^4$$

$$Q \propto 256 \times (D_1)^4$$

This indicates that flow rate increases 256 times when the diameter of the pipe increases four times. This is only applicable when flows are laminar as given in the question.

Correct Answer is (D)

SOLUTION 3.9
Using the *Joukowsky's* equation of Section 6.3.7 of the *NCEES Handbook version 2.0*, start with determining the critical time t_c which does not cause water hammer.

$$t_c = \frac{2L}{a} = \frac{2 \times 3{,}000\ ft}{3{,}200\ ft/sec} = 1.875\ sec$$

Since the actual time of closure is greater than t_c, equation of Section 6.3.7.3 is used as follows:

$$\Delta p = \rho a \Delta v \frac{t_c}{t_{act}}$$

Where:

$$\Delta v = \left|0 - \frac{15\ ft^3/sec}{\pi \times 1^2\ ft}\right| = 4.8\ ft/sec$$

$$\Delta p = 1.94\ \frac{lbf \cdot sec^2}{ft^4} \times 3{,}200\ \frac{ft}{sec} \times 4.8\ \frac{ft}{sec} \times \frac{1.875}{4.0}$$

$$= 13{,}968\ lbf/ft^2$$

SOLUTION 3.10
The *NCEES Handbook*, Section 6.3 Closed Conduit Flow and Pumps can be used to solve this problem. *Darcy-Weisbach* equation for head losses due to flow – Section 6.2.3.1 and Minor Losses in Pipe Fittings, Contractions and Expansions – Section 6.3.3 are both used in this question.

Since the network is horizontal, the head increase required by the pump equals to the minor and major losses $h_{f,total}$.

Head loss due to flow:

Using *Darcy-Weisbach* equation for head losses:

$$h_{f,flow} = f \frac{L}{D} \frac{v^2}{2g}$$

f is a function of R_e and the relative roughness $\left(\frac{\varepsilon}{D}\right)$ taken from the *Moody, Darcy, or Stanton* Friction Factor Diagram page 312 of the *NCEES Handbook version 2.0*.

v in the denominator of *Reynolds* number equation (shown below) is the kinematic viscosity taken from the physical properties of water table Section 6.2.1.6 as $0.93 \times 10^{-5}\ ft^2/sec$.

v in the numerator is velocity in the pipe

$$v_{10\ cfs} = \frac{Q}{A} = \frac{10\ ft^3/sec}{\pi\ (0.5\ ft)^2} = 12.7\ ft/sec$$

$$v_{5\ cfs} = \frac{Q}{A} = \frac{5\ ft^3/sec}{\pi\ (0.5\ ft)^2} = 6.4\ ft/sec$$

$$R_{e,v_{10\ cfs}} = \frac{vD}{v}$$

$$= \frac{12.7\ ft/sec \times 12\ in \times \frac{1\ ft}{12\ in}}{0.93 \times 10^{-5}\ ft^2/sec}$$

$$= 1.37 \times 10^6$$

$$\frac{\varepsilon}{D} = \frac{0.0006}{1\ ft} = 0.0006$$

Substitute $R_{e,v_{10\ cfs}}$ and $\frac{\varepsilon}{D}$ in *Moody, Darcy, or Stanton* Friction Factor Diagram page 312 generates a friction of $f \cong 0.018$

$$h_{f,flow\ @10\ cfs} = f\frac{L}{D}\frac{v^2}{2g}$$

$$= 0.018 \times \frac{(300\ ft)}{1\ ft}\frac{\left(12.7\frac{ft}{sec}\right)^2}{2\times 32.174\frac{ft}{sec^2}}$$

$$= 13.5\ ft$$

$$R_{e,v_{5\ cfs}} = \frac{vD}{v}$$

$$= \frac{6.4\ ft/sec \times 12\ in \times \frac{1\ ft}{12\ in}}{0.93\times 10^{-5}\ ft^2/sec}$$

$$= 0.7 \times 10^6$$

$$\rightarrow f \cong 0.018$$

$$h_{f,flow\ @5\ cfs} = f\frac{L}{D}\frac{v^2}{2g}$$

$$= 0.018 \times \frac{(400\ ft)}{1\ ft}\frac{\left(6.4\frac{ft}{sec}\right)^2}{2\times 32.174\frac{ft}{sec^2}}$$

$$= 4.6\ ft$$

Head loss due to pipe fittings:

$$h_{f,fittings} = C\frac{v^2}{2g}$$

Where C is velocity and fitting type dependent. There are four 45° elbows at flow $10\ cfs$ and another four at flow of $5\ cfs$. Also, T connection losses are taken into account as follows:

$$h_{f,fittings} = C_{total}\frac{v_{10\ cfs}^2}{2g} + C_{total}\frac{v_{5\ cfs}^2}{2g}$$

$$= (4 \times 0.4) \times \frac{\left(12.7\frac{ft}{sec}\right)^2}{2\times 32.174\frac{ft}{sec^2}}$$

$$+ (4 \times 0.4 + 2 \times 0.6 + 2 \times 1.8) \times \frac{\left(6.4\frac{ft}{sec}\right)^2}{2\times 32.174\frac{ft}{sec^2}}$$

$$= 8.1\ ft$$

Total head losses (i.e., required pump head) is as follows:

$$= 13.5 + 4.6 + 8.1 = 26.2\ ft$$

Correct Answer is (A)

SOLUTION 3.11

The *NCEES Handbook,* Section 6.3.6 Flow Through a Packed Bed is used to solve this question, in which case, *Ergun* equation is applied as shown below:

$$\frac{\Delta P}{L} = \frac{150v_o u(1-\varepsilon)^2}{k\phi_s^2 D_P^2 \varepsilon^3} + \frac{1.75\rho v_o^2 (1-\varepsilon)}{k\phi_s D_P \varepsilon^3}$$

In the above equation, it is desired to find the linear velocity, based upon which the cross-sectional area can be determined using the given flow rate.

ΔP is the pressure loss and is measured in (lbf/in^2), $40\ kPa = 5.8\ psi$

L is bed depth in $feet = 2\ ft$

u is the water's dynamic viscosity taken from Section 6.2.1.6 of the handbook as $0.00089\ Pa.s$ for water at temperature of $25^o\ C = 0.89\ Centipoise\ (cP)$. The equation desires this unit to be in $lbm/hr.ft$:

$$= 0.89\ cP \times 2.42 = 2.15\ lbm/hr.ft$$

ϕ_s is the shape factor given as 1.0

D_P is particle diameter $= 0.01\ ft$

$$\frac{5.8}{2} = \frac{150 \times v_o \times 2.15 \times (1-0.45)^2}{144 \times 1.0^2 \times (0.01)^2 \times 0.45^3}$$

$$+ \frac{1.75 \times 1.94 \times v_o^2 \times (1-0.45)}{144 \times 1.0 \times 0.01 \times 0.45^3}$$

$$2.9 = 74{,}345.6\, v_o + 14.2 v_o^2$$

$$v_o^2 + 5{,}225\, v_o - 0.2 = 0$$

$$v_o = 5{,}225\ ft/hr = 1.45\ ft/sec$$

$$A = \frac{Q}{V} = \frac{5\ ft^3/sec}{1.45\ ft/sec} = 3.44\ ft^2$$

Correct Answer is (C)

SOLUTION 3.12
Reference is made to the *NCEES Handbook* Section 6.3.8.4 Fluid Power Equations.

Start with determining the total head along with all the other losses as follows:

$$H = 37\ ft + 12.5\ ft = 49.5\ ft$$

Pump (brake) power is calculated as follows:

$$\dot{W} = \frac{\rho g H Q}{\eta_{pump}}$$

$$= \frac{1.94\,\frac{lbf.sec^2}{ft^4} \times 32.174\,\frac{ft}{sec^2} \times 49.5\ ft \times 2\ MGD \times 1.547\,\frac{ft^3/sec}{MGD}}{0.75}$$

$$= 12{,}745.9\ ft.lbf/sec\ (= 23.2\ hp)$$

Correct Answer is (A)

SOLUTION 3.13
Scaling and Affinity Laws of Section 6.3.8.5 of the *NCEES Handbook* are used in this question as follows:

$$\frac{Q_2}{N_2 D_2^{\,3}} = \frac{Q_1}{N_1 D_1^{\,3}}$$

$$Q_2 = \frac{N_2 D_2^{\,3}}{N_1 D_1^{\,3}} \times Q_1$$

$$= \frac{2{,}250\ rpm \times (12\ in)^3}{1{,}750\ rpm \times (10\ in)^3} \times 0.5\ MGD$$

$$= 1.11\ MGD$$

Correct Answer is (D)

SOLUTION 3.14
This problem is resolved by using cavitation equations provided in Section 6.3.8.6 of the *NCEES Handbook*. Although the problem is resolved by the direct application of an equation, will start with a brief explanation and a sketch.

Whenever a pump is positioned above a supply reservoir, the water pressure in the suction line will become lower than the atmospheric pressure and risk of cavitation will occur which will damage the impellers over time. To prevent this from occurring, the total suction head of the pump $H_{S,total}$ shall be determined such that it is less than the atmospheric head and the vapor head.

$$H_{S,total} < \frac{P_{atm}}{\rho g} - \frac{P_{vapor}}{\rho g}$$

$$< H_{pa} - H_{vp}$$

Net Positive Pressure Suction Head (NPSH) represents a pressure drop due to the action of the impellers – it also represents the minimum suction head required by the pump to operate without cavitation. NPSH is added to the losses at the suction side $\sum h_L$ and used to determine the static suction head of the liquid H_S as follows:

$$H_S = (H_{pa} - H_{vp}) - \left(NPSH_r + \sum h_L\right)$$

This equation should be modified to account for the pressure drop due to accelerating the flow from a stationary position at suction into the inlet pipe ($v^2/2g$) as follows:

$$H_S = (H_{pa} - H_{vp}) - \left(\frac{v^2_{inlet}}{2g} + \sigma h_p + \sum h_L\right)$$

Where h_p is the head added by the pump and is inclusive of all the losses:
$$= 50\,ft + 4\,ft + 27.5\,ft = 81.5\,ft$$

And $\sum h_L$ is the total friction losses in the suction line only $= 4\,ft$

Section 6.2.1.6 is used to calculate the atmospheric and vapor heads as follows:

$$H_{vp} = \frac{P_{vapor}}{\gamma} = \frac{0.51\frac{lb}{in^2} \times 144\frac{in^2}{ft^2}}{62.22\frac{lb}{ft^3}} = 1.2\,ft$$

$$H_{pa} = \frac{P_{atm}}{\gamma} = \frac{14.696\frac{lb}{in^2} \times 144\frac{in^2}{ft^2}}{62.22\frac{lb}{ft^3}} = 34.0\,ft$$

The pressure drop due to the acceleration of flow from its stationary position is:

$$\frac{v^2_{inlet}}{2g} = \frac{\left(\frac{10\,ft^3/sec}{\pi \times 0.5^2\,ft^2}\right)^2}{2 \times 32.174\,ft/sec^2} = 2.5\,ft$$

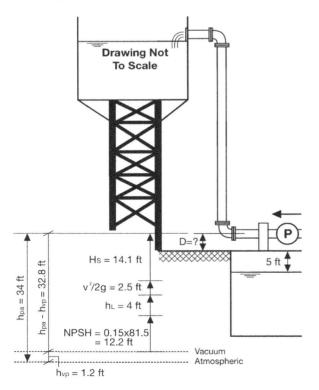

$$H_S = (34.0 - 1.2) - (2.5 + 0.15 \times 81.5 + 4)$$
$$= 14.1\,ft$$

$$D = 14.1 - 5 = 9.1\,ft$$

Correct Answer is (A)

SOLUTION 3.15

The *Hardy-Cross* method is not explained in the *NCEES Handbook* and deriving the correction factor requires successive iterations. The correction factor was given in the question and all what is left is to apply it to the estimated flows as shown in the following table were counterclockwise flows are depicted as negative:

Pipe	Q_o	ΔQ	Q_{new}
AC	+4	+1.44	+5.44
CB	0	+1.44	+1.44
AB	−6	+1.44	−4.56

This table is portrayed in the below pipe network diagram to which it shows that the flow from point B $= 1.44 + 4.56 = 6\,cfs$

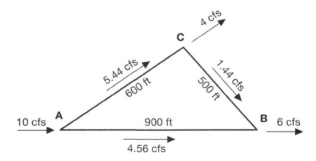

Before concluding with this answer, we shall verify option A by confirming that the summation of the loop head losses equals to zero. See below:

$$h_f = f\frac{Lv^2}{D \times 2g}$$

$$= f \frac{L \left(\frac{Q}{A}\right)^2}{D \times 2g}$$

$$= f \frac{L \left(\frac{Q}{\pi \times (0.5D)^2}\right)^2}{D \times 2g}$$

$$= \left(\frac{8fL}{g\pi^2 D^5}\right) Q^2$$

$$h_{f,AC} = \frac{8 \times 0.02 \times 600}{32.174 \times \pi^2 \times (1)^5} \times (5.44)^2$$

$$= +8.95$$

$$h_{f,CB} = \frac{8 \times 0.02 \times 500}{32.174 \times \pi^2 \times (1)^5} \times (1.44)^2$$

$$= +0.52$$

$$h_{f,AB} = -\frac{8 \times 0.02 \times 900}{32.174 \times \pi^2 \times (1)^5} \times (4.56)^2$$

$$= -9.43$$

$$\Sigma h_f = h_{f,AC} + h_{f,CB} + h_{f,AB}$$

$$= 8.95 + 0.52 - 9.43$$

$$= 0$$

This means that the correction factor, and the new flows, are all accurate, which rules out option A.

Correct Answer is (B)

SOLUTION 3.16

This is a parallel pipe network with two pipes, and hence two equations can be used to solve it – i.e., one that maintains head losses' continuity and the other maintains flow continuity. Check Section 6.3.10.1 and 2 of the *NCEES Handbook* for more details on those equations.

Flows continuity:

$$Q_{ACB} = 5 \times Q_{AB}$$

Head losses continuity:

$$f \frac{L_{AB} v_{AB}^2}{D_{AB} \times 2g} = f \frac{L_{ACB} v_{ACB}^2}{D_{ACB} \times 2g}$$

$$\frac{L_{AB} v_{AB}^2}{D_{AB}} = \frac{L_{ACB} v_{ACB}^2}{D_{ACB}}$$

$$\frac{L_{AB} \left(\frac{Q_{AB}}{A_{AB}}\right)^2}{D_{AB}} = \frac{L_{ACB} \left(\frac{Q_{ACB}}{A_{ACB}}\right)^2}{D_{ACB}}$$

$$\frac{L_{AB} \left(\frac{Q_{AB}}{0.25\pi D_{AB}^2}\right)^2}{D_{AB}} = \frac{L_{ACB} \left(\frac{Q_{ACB}}{0.25\pi D_{ACB}^2}\right)^2}{D_{ACB}}$$

$$= \frac{L_{ACB} \left(\frac{5 \times Q_{AB}}{0.25\pi D_{ACB}^2}\right)^2}{D_{ACB}}$$

$$\rightarrow \frac{L_{AB}}{D_{AB}^5} = \frac{25 \times L_{ACB}}{D_{ACB}^5}$$

$$D_{AB} = \left(\frac{L_{AB}}{25 \times L_{ACB}}\right)^{1/5} D_{AB}$$

$$= \left(\frac{900 \, ft}{25 \times 1,100 \, ft}\right)^{1/5} \times 12 \, in$$

$$\cong 6.0 \, in$$

Correct Answer is (B)

SOLUTION 3.17

Manning equation is used to solve this question.

$$Q = \frac{1.486}{n} A R_H^{2/3} S^{1/2}$$

Q is the discharge or flow rate (cfs).

A is the cross-sectional area of the flow (ft^2).

R_H is the hydraulic radius, this can be calculated by dividing the area of the flow (A) by the wetted Perimeter (P), or by using the tables provided in the *NCEES Handbook* Section 6.4.5.4.

S slope (ft/ft).

n is the *Manning* roughness coefficient found in the *NCEES Handbook version 2.0* page 346 for concrete lined channel '0.015'.

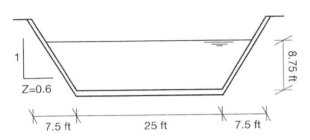

Using equations provided by *NCEES Handbook* and sourced by Chow (1959):

$$R_H = \frac{(b+zy)y}{b+2y\sqrt{1+z^2}}$$

$$= \frac{(25+0.6\times 8.75)\times 8.75}{25+2\times 8.75\times\sqrt{1+0.6^2}}$$

$$= 5.83\ ft$$

$A = (b + zy)y$
$= (25 + 0.6 \times 8.75) \times 8.75 = 264.7\ ft^2$

$Q = \frac{1.486}{0.015} \times 264.7 \times 5.83^{2/3} \times 0.02^{1/2}$
$= 12{,}012.75\ ft^3/sec\ (\cong 7{,}765\ MGD)$

Correct Answer is (B)

SOLUTION 3.18
Reynolds number R_e is calculated as follows:

$$R_e = \frac{vR_H}{v}$$

v is the mean velocity of the flow (given as 0.3 ft/sec).

v is the kinematic viscosity for water which equals to $1.059 \times 10^{-5}\ ft^2/sec$ for water at $70°\ F$ (See Section 6.2.1.6 of the handbook).

R_H is the hydraulic radius and equals to A/P which is calculated as follows for such a channel cross section (refer to the *NCEES Handbook* Section 6.4.5.4):

$$R_H = \frac{zy}{2\sqrt{1+z^2}}$$

$$= \frac{0.75 \times 3}{2\sqrt{1+0.75^2}}$$

$$= 0.9\ ft$$

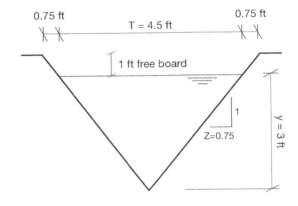

$$R_e = \frac{0.3 \times 0.9}{1.059 \times 10^{-5}}$$

$$\cong 25{,}496$$

Correct Answer is (D)

SOLUTION 3.19
The curb and gutter equation from Section 6.4.10.3 of the *NCEES Handbook* is used to solve this problem.

$$Q = \left(\frac{0.56}{n}\right) \frac{S_L^{0.5}}{S_x} d^{8/3}$$

Q is the discharge or flow rate (cfs).

S_x is the cross slope in which case is $0.12 \, ft/ft$ and S_L is the longitudinal slope being $0.01 \, ft/ft$.

n is the *Manning* roughness coefficient found in the *NCEES Handbook version 2.0* for concrete as '0.013' (also given in the question).

$$Q = \left(\frac{0.56}{0.013}\right) \frac{0.01^{0.5}}{0.12} \, 0.5^{8/3}$$
$$= 5.7 \, cfs$$

Correct Answer is (A)

SOLUTION 3.20
Reference is made to the *NCEES Handbook version 2.0* page 370, Chart 1B – Headwater Depth For Concrete Pipe Culverts with Inlet Control. The chart is pasted with permission from the Federal Highway Administration here for ease of reference.

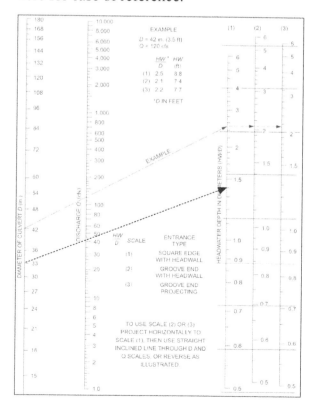

A straight line is drawn from the first scale that represents the diameter of the culvert.

The line starts at a diameter of $33 \, in$ as given in the question, and it passes through a flow of $40 \, cfs$ in the following scale which determines the value of HW/D in the first of the three scales to the rightmost of the graph, which represents a square edge with the headwall:

$$HW/D = 1.4$$

With a diameter $D = 33 \, in$:

$$HW = 1.4 \times 33 = 46.2 \, in \, (= 3.85 \, ft)$$

$$Elevation \, of \, HW = 317 + 3.85 = 320.85 \, ft$$

Correct Answer is (B)

SOLUTION 3.21
Reference is made to the *NCEES Handbook version 2.0* page 373, Chart 5B – Head for Concrete Pipe Culverts Flowing Full. The chart is pasted with permission from the Federal Highway Administration on the following page for ease of reference.

The entrance loss coefficient K_e shall be determined first for use in the said chart. This coefficient can be taken from the *NCEES Handbook* page 369 – taken as '0.5' for projected concrete pipes with a cut/square-edge end.

Using the graph copied below, start with the dotted line shown. The dotted line connects the diameter scale at $33 \, in$ and the length scale of $K_e = 0.5$ at $200 \, ft$. An **x** is marked on the turning line for later use.

A solid line/arrow is then extended starting from a discharge of $40 \, cfs$, passing through the **x** marked previously, and points towards the last scale (the head) which reads $2.1\ ft$ in this case.

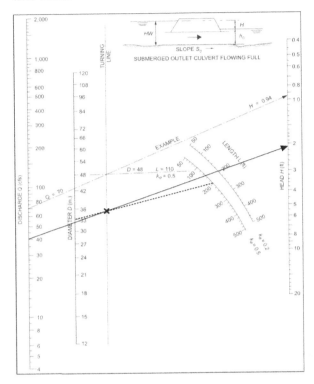

The above information is valid as long as $0.5 < HW/D < 3.0$. This information can be verified later with the help of the following sketch:

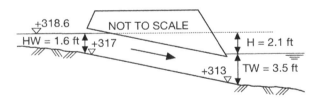

The requested water elevation at inlet can be determined graphically, considering the level at the inlet (+317) and the level at the outlet (+313), or by using the following equation:

$$HW = TW + H - LS$$

TW is the tailwater depth ($3.5\ ft$), H is the head calculated previously ($2.1\ ft$) and LS is the drop in elevation ($4\ ft$).

$$HW = 3.5 + 2.1 - 4 = 1.6\ ft$$

$$\frac{HW}{D} = \frac{1.6\ ft}{33\ in \times \frac{1\ ft}{12\ in}} = 0.58 \rightarrow ok$$

$$Elevation\ of\ HW = 317 + 1.6 = 318.6\ ft$$

Correct Answer is (A)

SOLUTION 3.22

The *NCEES Handbook version 2.0*, Section 6.4.3.1 *Froude* number, and Section 6.4.8.1 Depths and Flows, are both referred to in this solution.

Calculate *Froude* number for use in subsequent equations per Section 6.4.3.1 as follows:

$$Fr_1 = \frac{v}{\sqrt{gy_1}}$$

$$v = \frac{Q}{A} = \frac{80\ ft^3/sec}{8\ ft \times 1\ ft} = 10\ ft/sec$$

$$g = 32.174\ ft/sec^2$$

$$Fr_1 = \frac{10\ ft/sec}{\sqrt{32.174\ ft/sec^2 \times 1\ ft}} = 1.76\ (*)$$

$$\frac{y_2}{y_1} = \frac{1}{2}\left(\sqrt{1 + 8Fr_1^2} - 1\right)$$

$$\frac{y_2}{y_1} = \frac{1}{2}\left(\sqrt{1 + 8 \times (1.76)^2} - 1\right) = 2.0$$

$$y_2 = 2.0 \times 1 = 2.0\ ft$$

Another shorter method can be used by applying the following equation found on page 354 of the *NCEES Handbook version 2.0*:

$$y_2 = -\frac{1}{2}y_1 + \sqrt{\frac{2v_1^2 y_1}{g} + \frac{y_1^2}{4}}$$

$$= -\frac{1}{2}(1) + \sqrt{\frac{2(10)^2(1)}{32.174} + \frac{(1)^2}{4}}$$

$$= 2.0 \, ft$$

Correct Answer is (C)

(*) In reference to Section 6.4.8.2, a flow with *Froude* number which falls between $'1.0'$ and $'1.7'$ generates low energy jumps.

SOLUTION 3.23
The *NCEES Handbook version 2.0*, Section 6.4.8.6 Energy Loss in Horizontal Hydraulic Jump is referred to in this solution.

With the lack of information on velocity, the following equation can be used:

$$\Delta E = \frac{(y_2 - y_1)^3}{4 y_1 y_2} = \frac{(4-1)^3}{4(4)(1)} = 1.7 \, ft$$

Correct Answer is (B)

SOLUTION 3.24
The *Chezy* equation of Section 6.4.5.2 from the *NCEES Handbook* is used to solve this problem.

$$Q = CA\sqrt{R_H S}$$

Where R_H is the hydraulic radius for this channel which is calculated using the area of flow divided by the wetted perimeter as follows:

$$A = \left(\frac{\pi \times (3 \, ft)^2}{2}\right) + (6 \, ft \times 1.5 \, ft) = 23.1 \, ft^2$$

$$P = (\pi \times 3 \, ft) + (1.5 \, ft + 1.5 \, ft) = 12.4 \, ft$$

$$R_H = \frac{A}{P} = \frac{23.1}{12.4} = 1.86 \, ft$$

$$Q = 55 \times 23.1 \times \sqrt{1.86 \times 0.01}$$

$$= 173.3 \, cfs \, (= 112 \, MGD)$$

Correct Answer is (B)

SOLUTION 3.25
Section 6.4.5.3 of the *NCEES Handbook* is used to solve this problem.

Start with determining the minimum pipe diameter that can carry this flow full:

$$D = \left[\frac{2.16 Q n}{\sqrt{S}}\right]^{3/8}$$

$$= \left[\frac{2.16 \times 10 \, cfs \times 0.015}{\sqrt{0.005}}\right]^{3/8}$$

$$= 1.77 \, ft \, (= 21.2 \, in) < 30 \, in$$

Which means the pipe is not flowing full, and this excludes the first option.

Calculate full flow rate for a 30 *in* pipe using *Manning* equation as follows:

$$Q_{full} = \frac{1.486}{n} A R_H^{2/3} S^{1/2}$$

$$= \frac{1.486}{n} A \left(\frac{A}{P}\right)^{2/3} S^{1/2}$$

$$A = \pi \times \left(\frac{15}{12}\right)^2 = 4.9 \, ft^2$$

$$P = \pi \times \frac{30}{12} = 7.85 \, ft$$

$$Q_{full} = \frac{1.486}{0.015} \times 4.9 \times \left(\frac{4.9}{7.85}\right)^{2/3} 0.005^{1/2}$$

$$= 25.1 \, cfs$$

Calculate the ratio of the required flow to full flow and plug it into the chart provided in the same section of the handbook to determine the ratio of depth to the pipe diameter (chart provided by the ASCE – copied below for ease of reference):

$$\frac{Q}{Q_{full}} = \frac{10}{25.1} = 0.4$$

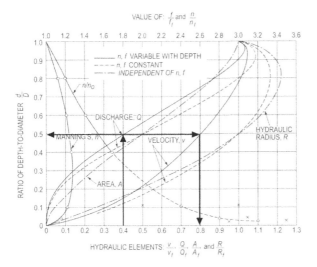

From the chart (*):

$$\frac{d}{D} = 0.5$$

$$\rightarrow d = 0.5 \times 30 = 15.0 \ in$$

From the same chart, it can be observed that:

$$\frac{v}{v_{full}} = 0.8$$

$$\rightarrow v = 0.8 \times \frac{25.1 \ cfs}{4.9 \ ft^2} = 4.1 \ ft/sec$$

Although this information was not requested in the question, velocity in such cases shall be kept within certain limits, hence calculated here for demonstration purposes only.

Correct Answer is (D)

(*) For clarity, the chart depicts two lines: one dotted representing a constant *Manning* coefficient, and a solid line representing a variable *Manning* coefficient. While the *Manning* roughness coefficient does indeed vary with the depth of flow, for simplicity, we often assume it to be constant. Whilst having access to this information using this chart, the variable/solid line is used to solve this problem.

SOLUTION 3.26

Section 6.4.10 of the *NCEES Handbook* is used to solve this problem.

Per the sketch provided in the question, the classification of this flow is a type 4 (Submerged Outlet), and the following flow equation should be used:

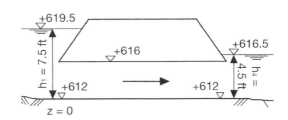

$$Q = CA_o \sqrt{\frac{2g(h_1 - h_4)}{1 + \frac{29C^2 n^2 L}{R_o^{4/3}}}}$$

$$= 0.95 \times \left(\frac{4^2 \times \pi}{4}\right)$$

$$\times \sqrt{\frac{2 \times 32.174 \times (7.5 - 4.5)}{1 + \frac{29 \times 0.95^2 \times 0.015^2 \times 75}{(0.25 \times 4)^{4/3}}}}$$

$$= 138.1 \ cfs$$

Where R_o is the hydraulic radius which equals to $D/4$ for circular full pipes.

Correct Answer is (A)

SOLUTION 3.27

The critical depth y_c is used along with the *Manning* equation to find the critical slope.

The critical depth for a rectangular channel can be calculated using equations in Section 6.4.3.2 of the *NCEES Handbook* as follows:

$$y_c = \left(\frac{Q^2}{b^2 g}\right)^{1/3}$$

$$= \left(\frac{65^2}{5^2 \times 32.174}\right)^{1/3}$$

$$= 1.74 \, ft$$

The *Manning* equation:

$$Q = \frac{1.486}{n} A R_H^{2/3} S^{1/2}$$

$$\rightarrow S = \left(\frac{Qn}{1.486 A R_H^{2/3}}\right)^2$$

$A = 5 \, ft \times 1.74 \, ft = 8.7 \, ft^2$

$P = 5 \, ft + 2 \times 1.74 \, ft = 8.48 \, ft$

$R_H = \frac{A}{P} = \frac{8.7}{8.48} = 1.03 \, ft$

$$S = \left(\frac{65 \times 0.013}{1.486 \times 8.7 \times 1.03^{2/3}}\right)^2$$

$$= 0.004 \, (= 0.4\%)$$

Correct Answer is (A)

SOLUTION 3.28

The hydraulic radius is the area of flow divided by the wetter perimeter, and water to water overlap is not considered part of the wetted permitter:

$$R_H = \frac{A}{P} = \frac{5 \times 7}{5 + 7 + 3} = 2.33 \, ft$$

Correct Answer is (B)

SOLUTION 3.29

This problem requires the direct application of the *Manning* equation to find out the depth, however, since the top width of water changes with flow depth, few considerations shall be taken into account as follows:

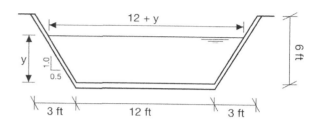

$A = y \times \left(\frac{(12+y)+(12)}{2}\right) = 0.5y(24 + y)$

$P = 12 + 2 \times \sqrt{y^2 + (0.5y)^2} = 12 + \sqrt{5y^2}$

$R_H = \frac{A}{P} = \frac{0.5y(24+y)}{12+\sqrt{5y^2}}$

$$Q = \frac{1.486}{n} A R_H^{2/3} S^{1/2}$$

$$80 = \frac{1.486}{0.013} \times [0.5y(24 + y)]$$

$$\times \left[\frac{0.5y(24+y)}{12+\sqrt{5y^2}}\right]^{2/3}$$

$$\times (0.001)^{1/2}$$

$$22.1 = [0.5y(24+y)] \times \left[\frac{0.5y(24+y)}{12+\sqrt{5y^2}}\right]^{2/3}$$

At this point, attempting to solve this equation using complex numerical methods would be time-consuming. Instead, it is more efficient to use iterations – i.e., attempt the four options given in the question to assess which one provides the correct answer.

$$\rightarrow y = 1.5 \, ft$$

Correct Answer is (C)

📖 SOLUTION 3.30

Although this question can be resolved with the trained eye, a detailed solution with proper explanation is provided below. Additionally, we recommend reviewing the brief discussion towards the end of this solution before delving into the details here.

Gradually Varied Flow profiles are all presented in the *NCEES Handbook* Sections 6.4.7.1 and 6.4.9.

Calculate critical depth

$$y_c = \left(\frac{q^2}{g}\right)^{1/3}$$

$$= \left(\frac{40^2}{32.174}\right)^{\frac{1}{3}}$$

$$= 3.7 \, ft$$

Calculate normal depth

$$y_{n,S=0.02} = \left(\frac{0.025q}{S^{1/2}}\right)^{3/5}$$

$$= \left(\frac{0.025 \times 40}{0.02^{1/2}}\right)^{3/5}$$

$$= 3.2 \, ft < 3.7 \, ft \text{ (Steep Slope)}$$

$$y_{n,S=0.001} = \left(\frac{0.025q}{S^{1/2}}\right)^{3/5}$$

$$= \left(\frac{0.025 \times 40}{0.001^{1/2}}\right)^{3/5}$$

$$= 7.9 \, ft > 3.7 \, ft \text{ (Mild Slope)}$$

Sketch the two depths

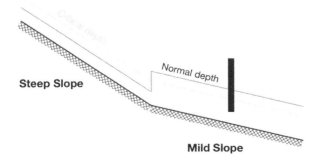

As evident from the depth calculations, the left-to-right slopes can be categorized into two segments: the first one is steep, while the subsequent slope is a mild slope. The critical depth is indicated by a light font and a light dotted line. It can be observed that the critical depth slope sits above the normal flow depth on the steep slope and below it on the mild slope.

Moving forward with the solution, we can consult the GVF (Gradually Varied Flow) curves in Section 6.7.4.1 of the handbook. Using these curves, the profile of how water will flow in the channel can be sketched as explained below.

Add Gradually Varied Flow GVF Profiles:

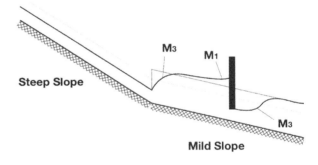

Starting left to right, the following GVF profiles have been identified: M_3, M_1 and finally M_3.

Profile M_3

As water approaches the mild slope, the normal depth (which is now forced into a shallow depth from the previous slope) will be located in zone 3 of the mild slope, and it has to go all the way up to its new normal depth. Because of this, the GVF will take the shape of M_3. S_3 curve could as well occur due to its proximity to the steep slope and there are methods to check if this could happen. Those methods however are beyond the scope of this exam.

Profile M_1

At this stage, and as water is bouncing against the weir, it will be forced into zone 1 of the mild slope, in which case GVF profile M_1 will occur.

Profile M_3

Continuing with the same logic of the previous profiles, the water is forced into zone 3 of the mild slope because of the sluice gate, in which case an M_3 profile is expected to occur with a hydraulic jump to regain its new normal position.

Correct Answer is (C)

📖 Gradually Varied Flows:

There are five distinct slope types, summarized in Section 6.4.6 of the *NCEES Handbook*: Mild, Steep, Critical, Horizontal, and Adverse. Each of these slopes generate different flow behaviors, resulting in a total of nearly three flow types per slope – i.e., a total of 15 unique flow scenarios. Among these, the mild and steep slopes hold particular significance, as they serve as foundational concepts from which other flow types can be derived.

Will explain the mild slope in here based upon which an understanding for the rest of flows can be accomplished – all flow types are provided in the handbook's Section 6.4.7.1.

The Mild Slope:

A mild slope occurs when the normal flow is above the critical flow. Below is a longitudinal section with a dotted line depicting where the normal flow is located compared to the critical flow for a mild slope – this relationship is reversed for a steep slope.

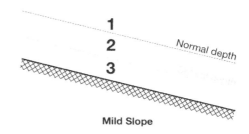

Consider the following observations from the above sketch:

Three Distinct Regions:

The sketch shows three distinct regions.

Normal Flow Interruption:

Under typical circumstances, flow within a channel maintains its normal depth until an interruption occurs. This interruption could be due to the presence of a weir, a gate, or a sudden change in slope.

Gradually Changing Flow Profile:

Interruptions lead to a gradual transformation of the flow profile. See below sketches for three possible profiles for the mild slope:

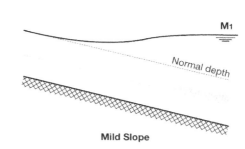

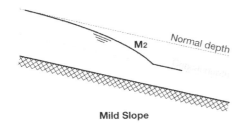

Mild Slope

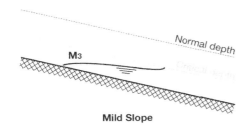

Mild Slope

These profiles are denoted by letters as shown in Section 6.4.7.1 of the handbook. For instance, 'M' signifies a mild slope, while 'S' represents a steep slope. Additionally, each profile is associated with a numerical zone designation: M_1, M_2 and M_3.

In a channel with a mild slope, when the flow is forced above the normal depth, specifically into zone 1, the flow takes shape of profile M_1. Same applies when the flow is forced into zone 2 where the flow gradually takes shape of profile M_2, and in a similar fashion for M_3.

The same logical framework applies to other slope types and their corresponding flow profiles.

Lastly, for better comprehension, consider sketching hypothetical slopes and obstructions. Try to predict how the flow will behave in these scenarios.

SOLUTION 3.31

A composite roughness coefficient can be calculated using the method presented in Section 6.4.10.2 of the *NCEES Handbook* page 362 of version 2.0 as follows:

$$n_c = \left[\frac{\sum_{i=1}^{G}(p_i n_i^{1.5})}{\sum P}\right]^{0.57}$$

Where n_i and p_i are the individual components of *Manning* coefficient and the wetted perimeter respectively as detailed in the following table:

Canal Component	Manning roughness coefficient	Wetted Perimeter length ft
1	0.011	3.35
2	0.03	2.5
3	0.01	2.24
4	0.45	5
5	0.01	1.12

$$n_c = \left[\frac{\begin{matrix}3.35 \times 0.011^{1.5} + 2.5 \times 0.03^{1.5} \\ + \\ 2.24 \times 0.01^{1.5} + 5.0 \times 0.45^{1.5} \\ + \\ 1.12 \times 0.01^{1.5}\end{matrix}}{3.35 + 2.5 + 2.24 + 5 + 1.12}\right]^{0.57}$$

$$= \left[\frac{1.53}{14.21}\right]^{0.57}$$

$$= 0.28$$

Correct Answer is (D)

SOLUTION 3.32

Section 6.4.10.2 of the *NCEES Handbook version 2.0* page 361 and the following energy equation is used:

$$H_u + LS + \frac{v_u^2}{2g} = H_d + \frac{v_d^2}{2g} + H_L$$

$H_u, LS\ \&\ H_d$ are defined on the following sketch:

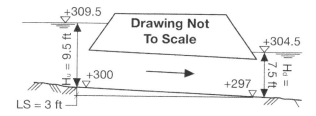

Total head losses are calculated using the friction head loss equation presented on page 362 of the *NCEES Handbook version 2.0*, and for the removal of doubt, this equation is inclusive of entrance losses:

$$H_L = H_f = \left(1 + k_e + \frac{K_u n^2 L}{R_H^{1.33}}\right)\frac{v^2}{2g}$$

$$= \left(1 + 0.5 + \frac{29 \times 0.015^2 \times 100}{(0.25 \times 6)^{1.33}}\right)\frac{15^2}{2 \times 32.174}$$

$$= 6.6\ ft$$

Substitute this information in the energy equation and solve for v_d:

$$v_d = \sqrt{2g\left(H_u + LS + \frac{v_u^2}{2g} - H_d - H_L\right)}$$

$$= \sqrt{2 \times 32.174\left(9.5 + 3 + \frac{15^2}{2 \times 32.174} - 7.5 - 6.6\right)}$$

$$= 11.1\ ft/sec$$

Correct Answer is (B)

PART III
Hydraulics

IV
HYDROLOGY, SURFACE WATER, GROUNDWATER & WELLS

Knowledge Areas Covered

SN	Knowledge Area
7	**Hydrology** A. Storm characteristics (e.g., storm frequency, rainfall measurement, distribution) B. Runoff analysis (e.g., rational and SCS/NRCS methods) C. Hydrograph development and applications, including synthetic hydrographs D. Rainfall intensity, duration, frequency, and probability of exceedance E. Time of concentration F. Rainfall and stream gauging stations G. Depletions (e.g., evaporation, detention, percolation, diversions) H. Stormwater management and treatment (e.g., detention and retention ponds, infiltration, swales, constructed wetlands)
8	**Groundwater and Wells** A. Aquifers B. Groundwater flow C. Well and drawdown analysis

PART IV
Hydrology

PROBLEM 4.1 *Watershed Rainfall Depth*

The following watershed has a total area of 0.31 *Acre* and is plotted to scale on a 10 *ft* × 10 *ft* grid.

Rain gauge stations 1, 2, 3 and 4 have been placed as shown and they measure the following rain depths:

Station 1 = 7.5 *in*

Station 2 = 5.5 *in*

Station 3 = 11.5 *in*

Station 4 = 7.5 *in*

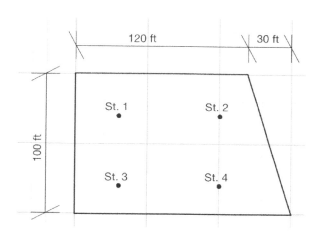

The average rainfall depth over the shown watershed using the *Thiessen* method is most nearly:

(A) 8.0 *in*

(B) 7.8 *in*

(C) 7.6 *in*

(D) 7.5 *in*

PROBLEM 4.2 *Precipitation Methods*

The most accurate method for averaging precipitation over an area is:

(A) The mathematical averaging method.

(B) The *Isohyetal* method.

(C) The *Thiessen* method.

(D) None of the above, each of those methods has a specific use and accuracy is irrelevant in this case.

PROBLEM 4.3 *Retention Pond Sizing*

The size of a retention pond in ft^3 situated in a sloped zone that has 50 *acres* of parks and cemeteries and the following data is most nearly:

- Outflow rate of 35 *cfs*.
- Water flows in and out the pond in 15 *minutes*.
- Average rainfall intensity is 3.5 *in/hr*.

(A) 7,875

(B) 131

(C) 39,375

(D) 31,500

PROBLEM 4.4 *Shallow Flow*

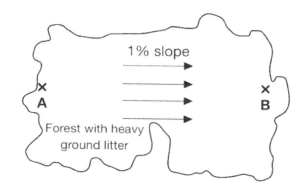

The above land is a forest characterized by heavy ground litter. This forest will be converted into short grass pasture. Land is sloped at a grade of $0.01\ ft/ft$.

The expected improvement in velocity of water flowing from point A to point B is most nearly:

(A) $0.74\ ft/sec$

(B) $0.25\ ft/sec$

(C) $0.50\ ft/sec$

(D) $1.00\ ft/sec$

PROBLEM 4.5 *Travel Time for Shallow Flow*

The below paved parking lot drains into the channel at its left side as shown. The parking has a slope of 0.5%.

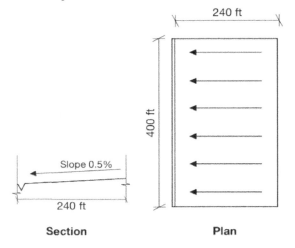

With a manning roughness of '0.011' and a 2-year 24-hour rainfall intensity of $2.0\ in/hr$. The travel time in minutes for a shallow flow over this plane to the channel is:

(A) 42.0 minutes

(B) 7.7 minutes

(C) 6.2 minutes

(D) 57.2 minutes

PROBLEM 4.6 *Retention Basin Design*

A slightly sloped $7\ acres$ piece of land that used to be an old cemetery is to be fully redeveloped as follows:

- 45% downtown areas
- 35% playgrounds
- 15% asphalt roads
- 5% concrete walkways

Given the rainfall intensity in this area is $2\ in/hr$, the depth of a $250\ ft \times 150\ ft$ retention basin to be constructed to store the excess runoff from this redevelopment for one day is most nearly:

(A) $14\ ft$

(B) $12\ ft$

(C) $16.5\ ft$

(D) $18\ ft$

PROBLEM 4.7 *Composite Curve Number*

The below table provides a summary of a watershed with several subcatchments each has a different curve number:

Description	Area (Acre)	Curve Number
Streets and roads	1.2	98
Residential	2	46
Loan & open space	1.5	45

Based on the above table, the rainfall excess for a $3\ in$ precipitation is most nearly:

(A) $1.3\ in$

(B) $0.3\ in$

(C) $7\ in$

(D) $3\ in$

PROBLEM 4.8 *10-Year Peak Flow*

The following data has been collected for a certain watershed:

- Area = 200 $hectares$
- Composite curve number $CN = 80$
- Time of Concentration = 1.2 hr
- Rain type is Type II
- 10-year rainfall of 7.4 in

Based on the above information, the 10-year peak flow using the SCS flow method is most nearly:

(A) 755 cfs

(B) 3,103 cfs

(C) 2,515 cfs

(D) 1,252 cfs

PROBLEM 4.9 *Hydrograph Excess Rain*

A triangular shaped hydrograph with a time duration of 120 $hours$ that runs over a 15 $mile^2$ catchment, 2 in initial abstraction and infiltration, and a peak flow of 800 cfs that occurs after 70 $hours$ form its start, has a rainfall excess of:

(A) 1 in

(B) 2 in

(C) 3 in

(D) 5 in

PROBLEM 4.10 *Hydrograph Development*

A 7.44 $mile^2$ watershed, with a time of concentration of 1.5 $hours$, has the following volume of excess rain when using a triangular unit hydrograph with a peak discharge:

(A) $18 \times 10^6 \ ft^3$

(B) $35 \times 10^6 \ ft^3$

(C) $40 \times 10^6 \ ft^3$

(D) $80 \times 10^6 \ ft^3$

PROBLEM 4.11 *Hydrograph Shape*

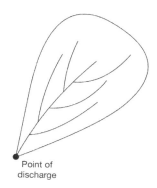

the best hydrograph that matches the above watershed would be the following:

(A) Option A

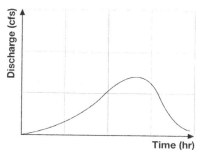

(B) Option B

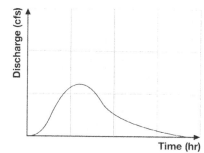

(C) Option C

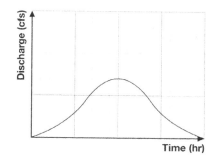

(D) Option D

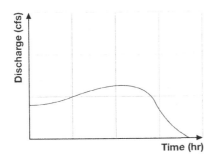

PROBLEM 4.12 *Rainfall Intensity*
Using the National Weather Service IDF equations, a storm event with a duration of 10 *hours* is expected to have the following design rainfall intensity:

(A) 0.3 *in/hr*

(B) 6.0 *in/hr*

(C) 1.8 *in/hr*

(D) 0.6 *in/hr*

PROBLEM 4.13 *Rain Gauge Precipitation Estimation*
The following sketch and table represent precipitation data and lengths for each gauge station (sketch not to scale):

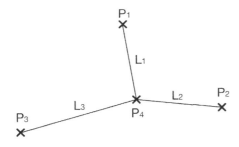

Gauge station P_1	3.0 *in*
Gauge station P_2	1.0 *in*
Gauge station P_3	7.0 *in*
Length L_1	5.0 *mile*
Length L_2	3.5 *mile*
Length L_3	10.0 *mile*

The expected precipitation at gauge P_4 is most nearly:

(A) 3.7 *in*

(B) 2.1 *in*

(C) 2.7 *in*

(D) 4.8 *in*

PROBLEM 4.14 *Exceedance Probability*
The return period of a flood event is 25 years. The probability that this event will be exceeded at least once in the coming 10 years is most nearly:

(A) 0.34

(B) 0.4

(C) 0.04

(D) 0.1

PROBLEM 4.15 *Horton Infiltration Capacity Curve*
The below is an infiltration capacity curve for a certain subcatchment:

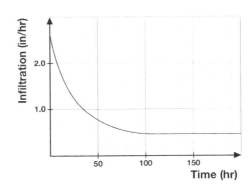

The following equation is the best representation of *Horton's* model for this curve:

(A) $f = 2.6e^{0.039t} - 0.45$

(B) $f = 2.15e^{-0.039t} + 2.6$

(C) $f = 2.15e^{0.039t} + 0.45$

(D) $f = 2.15e^{-0.039t} + 0.45$

PROBLEM 4.16 *Hydrologic Budget (1)*
A lake has an area of 800 *acre* with a measured inflow of 10 *cfs* and an outflow of 7 *cfs* along with and expected storage of 120 *acre.ft* receives precipitation of 9.5 *in* over the study year.

Ignoring any infiltration that could occur, groundwater flow or transpiration, the expected loss due to surface evaporation over the study period is most nearly:

(A) 3.5 *in*

(B) 8 *in*

(C) 40 *in*

(D) 23 *in*

PROBLEM 4.17 *Hydrologic Budget (2)*
Over a study period of 1 month, 50% of the 500 *acre* watershed discharge that is coming from precipitation is lost in evaporation, and the outflow at the discharge point is measured as 5 *cfs*.

Ignoring any other watershed losses (i.e., infiltration, transpiration, and groundwater flow), the expected precipitation over the study period is most nearly:

(A) 14 *in*

(B) 5 *in*

(C) 17 *in*

(D) 3 *in*

PROBLEM 4.18 *Unconfined Aquifer*
A 100 *ft* thick unconfined aquifer has a 12 *in* diameter well that pumps ground water from it at a rate of 65 *gpm* (gallon per minute).

Assuming the radius of influence is 450 *ft* and permeability is $4 \times 10^{-4} ft/sec$, the drawdown at the well is most nearly:

(A) 96 *ft*

(B) 97.5 *ft*

(C) 4 *ft*

(D) 2.5 *ft*

PROBLEM 4.19 *Weep Holes Flow*
The below is a retaining wall cross section with an elevated water table behind it, for which, weep holes were drilled at an interval of 10 *ft* at its bottom to lower it.

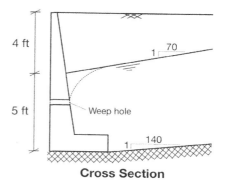

Cross Section

Given that permeability of the soil behind this retaining wall is 0.0004 *ft/sec*, the seepage through each weep hole is most nearly:

(A) 5.7 ft^3/day

(B) 12.5 ft^3/day

(C) 0.5 ft^3/day

(D) 25 ft^3/day

PROBLEM 4.20 *Confined Aquifer (1)*
A 75 *ft* thick confined aquifer that has attained an equilibrium condition after drawing water from it for a significant period has its well diameter at 12 *in* and a coefficient of permeability of 0.0005 *ft/sec*.

An observation well located 130 *ft* from the pumping well indicates that the slope of the piezometric surface is 0.03

Given that the aquifer thickness is 75 ft, the well discharge in gpm (gallon per minute), is most nearly:

(A) 412 gpm

(B) 312 gpm

(C) 800 gpm

(D) 590 gpm

PROBLEM 4.21 Confined Aquifer (2)
The below is a 50 ft confined aquifer pumping at a rate of 75 gpm (gallon per minute) and a drawdown of 65 ft occurs at its 12 in diameter well after a long time of pumping.

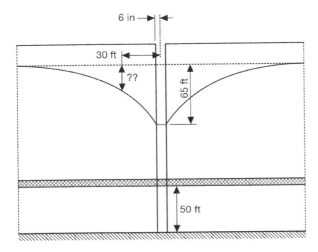

Given that permeability of the soil at this aquifer is 0.0004 ft/sec, the piezometric surface draw down 30 ft away from the well is most nearly:

(A) 45 ft

(B) 50 ft

(C) 55 ft

(D) 60 ft

PROBLEM 4.22 Composite Permeability
An aquifer is composed of four layers each with a different permeability as follows: 0.001 ft/sec, 0.0005 ft/sec, 0.0035 ft/sec, and 0.009 ft/sec with layers' thickness of 15 ft, 20 ft, 30 ft and 45 ft respectively.

The composite hydraulic conductivity for this aquifer is most nearly:

(A) 0.0035 ft/sec

(B) 0.0049 ft/sec

(C) 0.54 ft/sec

(D) 0.014 ft/sec

PROBLEM 4.23 Aquifer Porosity
An aquifer with a hydraulic conductivity of 85 ft/day with two wells set up 1.5 miles apart. The difference in the water level between the two wells is 9 ft, and it takes the water 3 years to move from the upper well to the lower well. Based on this information, the porosity of this aquifer is most nearly:

(A) 0.013

(B) 0.16

(C) 0.005

(D) 0.03

PROBLEM 4.24 Aquifer Volume
An aquifer runs over an area of 100 acre and has a porosity of 0.45 and specific retention of 0.2.

If the water level drops in this aquifer by 12 ft, this makes the original (bearing formation) depth of the aquifer most nearly:

(A) More information needed.

(B) 24 ft

(C) 48 ft

(D) 60 ft

SOLUTION 4.1

The *Thiessen* method is a weighing method with a weight assigned to each of the gauging stations. Straight lines (the solid lines) are drawn to connect all the stations as shown in the figure below. Perpendicular bisectors (the dotted lines) of those connecting lines are then extended to form polygons around each station.

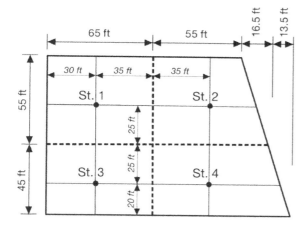

The area of each polygon is considered as the effective area per station.

A weighted average is then used to compute the required average rain depth.

St.	Effective area (A) ft^2	Gauge depth in	A × depth
1	3,575.00	7.5	26,812.50
2	3,478.75	5.5	19,133.13
3	2,925.00	11.5	33,637.50
4	3,521.25	7.5	26,409.38
	13,500.00		**105,992.50**

Average depth = 105,995.5/13,500 = 7.85 in

This method is referenced and explained in further detail in the *NCEES Handbook* Section 6.5.6.2 Thiessen Polygon Method.

Correct Answer is (B)

SOLUTION 4.2

The most accurate method for averaging precipitation over an area is the *Isohyetal* method. In this method, the amount of precipitation measured at each station is placed on the watershed map at gauges' locations. Contours of equal precipitation, which are called Isohyets, are then drawn.

The area between Isohyets is determined accurately using planimetry. Based on those measures, the average precipitation for each area between those Isohyets is estimated by taking the average of two Isohyets sharing a boundary.

These values are then multiplied by each area's percentage and added together to obtain the weighted average.

The *NCEES Handbook version 2.0,* Section 6.5.6 Rainfall Gauging Stations, has a detailed elaboration for this method, the *Thiessen* method, and the arithmetic averaging method as well.

Correct Answer is (B)

SOLUTION 4.3

Referring to Section 6.5.9.1 of the *NCEES Handbook*, the size of a retention pond using the rational method is calculated as follows – considering that the runoff coefficient C for sloped parks and cemeteries (check comment below the NCEES runoff table page 378 of version 2.0) is 0.25 per Section 6.5.2.1:

$$V_s = V_{in} - V_{out}$$
$$= (i \sum AC - Q_o) \times t$$
$$= \left(3.5 \frac{in}{hr} \times 50 \ acres \times 0.25 - 35 \ cfs\right) \times 15 \ min \times 60 \frac{sec}{min}$$
$$= 7,875 \ ft^3 \ (*)$$

Correct Answer is (A)

(*) Although the *NCEES Handbook version 2.0* identifies t as time in minutes, given the units consistency in this question, t is measured in seconds.

SOLUTION 4.4

The Water Velocity Versus Slope for Shallow Concentrated Flow diagram page 397 of the *NCEES Handbook version 2.0*, Section 6.5.5 Hydrograph Development and Applications can be used to solve this problem. The graph is pasted below for ease of reference.

A horizontal line is constructed from the y-axis at 0.01 which intersects with the desired two velocity graphs as shown:

$$v_{forest} = 0.25 \, ft/sec$$

$$v_{short\,grass} = 0.74 \, ft/sec$$

$$\Delta v = 0.74 - 0.25 = 0.49 \, ft/sec$$

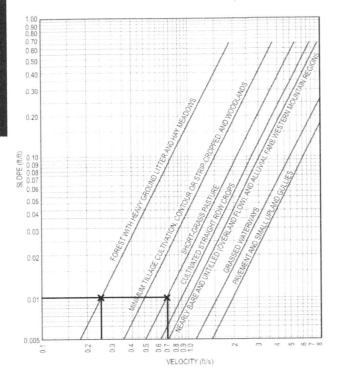

Correct Answer is (C)

SOLUTION 4.5

This is a sheet flow, which is a flow over plane surfaces at very shallow depths (about $0.1 \, ft$). Per *NCEES Handbook*, Section 6.5.4.2, sheet flow travel time in minutes over a flow length $L \, (ft)$, and a slope S measured in (ft/ft) is calculated as follows:

$$T_{ti} = \frac{K_u}{I^{0.4}} \left(\frac{nL}{\sqrt{S}}\right)^{0.6}$$

$$= \frac{0.933}{2.0^{0.4}} \left(\frac{0.011 \times 240 \, ft}{\sqrt{0.005}}\right)^{0.6} = 6.2 \, minutes$$

Correct Answer is (C)

SOLUTION 4.6

The *NCEES Handbook version 2.0*, Section 6.5.2 Runoff Analysis can be used. Given the input provided in the question, the best method to be implemented is the Rational Method.

$$Q = CIA$$

Q is the discharge in ft^3/sec, this value is calculated pre- and post- development, the difference of both shall be used to design the retention basin.

C is the runoff coefficient which can be obtained from the *NCEES Handbook*. Given the area is only slightly sloped, the lower range of C in the runoff table of page 378 of the handbook version 2.0 can be used.

I is rainfall intensity given in the question as $2 \, in/hr$, and A is area which is $7 \, acres$.

Pre-development:

For slightly sloped parks and cemeteries $C = 0.10$

$$Q_{pre} = CIA$$

$$= 0.1 \times 2 \ in/hr \times 7 \ acres$$
$$= 1.4 \ ft^3/sec$$

$$V_{pre} = Q_{pre} \times 1 \ day$$
$$= 1.4 \ ft^3/sec \times 86,400 \ sec \ per \ day$$
$$= 120,960 \ ft^3/day$$

Post-development:

Calculate the weighted runoff coefficient as follows:
- Downtown areas → $C = 0.70$
- Playgrounds → $C = 0.20$
- Asphalt roads → $C = 0.70$
- Concrete walkways → $C = 0.80$

$$C_w = 45\% \times 0.7 + 35\% \times 0.2$$
$$+ 15\% \times 0.7 + 5\% \times 0.8$$
$$= 0.53$$

$$Q_{post} = CIA$$
$$= 0.53 \times 2 \ in/hr \times 7 \ acres$$
$$= 7.42 \ ft^3/sec$$

$$V_{post} = Q_{post} \times 1 \ day$$
$$= 7.42 \ ft^3/sec \times 86,400 \ sec \ per \ day$$
$$= 641,088 \ ft^3/day$$

$$\Delta V = V_{post} - V_{pre}$$
$$= 641,088 - 120,960$$
$$= 520,128 \ ft^3$$

The depth of the rectangular pond is therefore calculated as follows:

$$d = \frac{520,128 \ ft^3}{250 \ ft \times 150 \ ft} = 13.87 \ ft$$

Correct Answer is (A)

SOLUTION 4.7

The *NCEES Handbook version 2.0*, Section 6.5.2.2 NRCS (SCS) Rainfall Runoff Method is used in this solution.

Start with determining the composite curve number using a weighted average method:

$$CN = \frac{98 \times 1.2 + 46 \times 2 + 45 \times 1.5}{1.2 + 2 + 1.5} = 59$$

Based on the composite curve number, the storage, or maximum basin retention, can be calculated as follows:

$$S = \frac{1,000}{CN} - 10 = \frac{1,000}{59} - 10 = 7.0 \ in$$

Based on this, the rain fall excess, or the runoff depth, is calculated as follows:

$$Q = \frac{(P - 0.2S)^2}{P + 0.8S} = \frac{(3 - 0.2 \times 7)^2}{3 + 0.8 \times 7} = 0.3 \ in$$

Correct Answer is (B)

SOLUTION 4.8

The *NCEES Handbook version 2.0*, Section 6.5.2.2 NRCS (SCS) Rainfall Runoff Method is used in this solution.

The storage, or maximum basin retention, can be calculated as follows:

$$S = \frac{1,000}{CN} - 10 = \frac{1,000}{80} - 10 = 2.5 \ in$$

The rainfall excess, or the runoff depth, is calculated as follows:

$$Q = \frac{(P - 0.2S)^2}{P + 0.8S} = \frac{(7.4 - 0.2 \times 2.5)^2}{7.4 + 0.8 \times 2.5} = 5.1 \ in$$

Use the above information to find I_a/P from page 386 of the handbook $I_a/P = 0.1$

Using this information, the following coefficients for SCS Peak Discharge Method can be collected from page 385 of the handbook for Type II rain:

$$C_o = 2.55323$$

$$C_1 = -0.61512$$

$$C_2 = -0.16403$$

Based on this information, the peak discharge can be calculated as follows:

$$q_p = q_u A_m Q$$

Where:

q_p is the required peak flow in cfs
A_m is basin area = $200\ hec\ (= 0.77\ mi^2)$

$$q_u = 10^{C_o + C_1 \log t_c + C_2 (\log t_c)^2}$$

$$= 10^{2.55323 - 0.61512 \log 1.2 - 0.16403 (\log 1.2)^2}$$

$$= 318.8\ \frac{cfs}{mi^2}/in$$

$$q_p = 318.8\ \frac{cfs}{mi^2}/in \times 0.77\ mi^2 \times 5.1\ in$$

$$= 1,251.9\ cfs$$

Correct Answer is (D)

📖 Rainfall Type II in the SCS method is one of the four synthetic 24-hour rainfall temporal distributions that are used to estimate design peak flows in small watersheds. The Type II distribution is characterized by a high peak intensity near the mid-point of the storm, with the maximum 5-minute rainfall depth equal to 11.7% of the total 24-hour depth. The Type II distribution is applicable to the interior regions of the USA and may not be suitable for other regions with different climatic conditions.

SOLUTION 4.9

Start with computing the volume of runoff by calculating the area under the triangular hydrograph as follows:

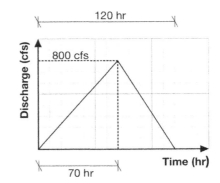

$$V = \frac{1}{2} \times \left(120\ hr \times \frac{3{,}600\ sec}{hr} \times 800\ \frac{ft^3}{sec}\right)$$

$$= 172.8 \times 10^6\ ft^3$$

Divide this volume by the area given in the question to determine the runoff depth:

$$Q = \frac{172.8 \times 10^6\ ft^3}{15\ mile^2 \times \frac{5{,}280^2\ ft^2}{mile^2}}$$

$$= 0.41\ ft\ (\approx 5\ in)$$

Remove the infiltration and initial abstraction of 2 in, the remaining excess rain = **3 in**

Correct Answer is (C)

SOLUTION 4.10

The SCS (NRCS) – the Natural Resources Conservation Services Unit Hydrograph Method – is used to solve this question (*). The NRCS Method is a widely known method and is presented in Section 6.5.5 of the *NCEES Handbook version 2.0*, and the specific equations are found in page 395.

The parameters that need to be determined are the peak discharge q_p, the time to peak t_p, and the duration. Those are calculated as follows:

$$q_p = \frac{K_p A_m Q_D}{t_p}$$

Where the time to peak t_p equals to two thirds the time of concentration:

$$t_p = \frac{2}{3} t_c = \frac{2}{3} \times 1.5 = 1 \ hr$$

$$q_p = \frac{484 \times 7.44 \ mi^2 \times 1 \ in}{1 \ hr} \approx 3{,}600 \ cfs$$

In order to calculate the area under the triangular unit hydrograph (i.e., the requested volume), the base of the triangle t_b is needed, and this can be determined using the unit hydrograph presented in page 395 as 2.7 hours.

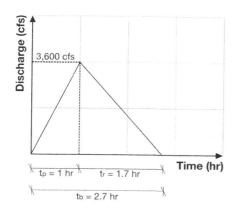

$$V = \frac{1}{2} \times \left(2.7 \ hr \times \frac{3{,}600 \ sec}{hr} \times 3{,}600 \ \frac{ft^3}{sec} \right)$$

$$= 17.5 \times 10^6 \ ft^3$$

Correct Answer is (A)

(*) The Snyder Synthetic Unit Hydrograph presented in page 394 can be used to solve this question as well, the base of the hydrograph in this case will equal to $T = 3 + \frac{t_p}{8} = 3.1 \ hours$ and with some approximations the volume will be slightly larger.

📖 SOLUTION 4.11

At the end of this solution, some information regarding the construction of a hydrograph is available.

When examining the provided watershed and dividing the area into zones, it can be observed that the shaded area closer to the point of discharge is quite narrow. This narrower area has a shorter time of concentration compared to the larger areas located farther away from the discharge point. Consequently, the flow at the beginning of a rainfall event will not reach its peak until the flow from the larger areas at the back of the catchment, which have the greatest time of concentration, arrives at the point of discharge.

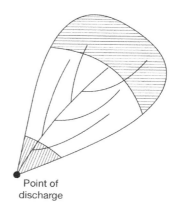

This means that the peak discharge point will be shifted to the right most of the hydrograph as shown below.

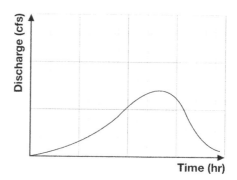

In summary, the interplay between different areas within the watershed, based on their time of concentration, allows us to predict how the hydrograph will evolve during a storm event.

Correct Answer is (A)

📖 Constructing a hydrograph

The process of constructing a hydrograph begins with a rainfall event. Heavy rain contributes to increased water flow in the watershed. The duration and intensity of the rainfall impact the hydrograph shape.

A hydrograph typically consists of two main components: (1) the rising limb which represents the increasing discharge as the rainwater flows into the watershed and it corresponds to the rising water level during the storm, and (2) the falling limb which corresponds to the decrease in flow as the rain subsides.

The peak discharge occurs when the flow in the water shed reaches its highest flow rate during the storm.

The time of concentration is the duration it takes for water to travel from the farthest point in the watershed to the outlet (where the hydrograph is measured), and because of that it influences the shape of the hydrograph. In which case, we need to understand the lag time, which is the is the delay between the peak rainfall and the peak discharge.

Factors affecting lag time include watershed size, slope, land use, shape of the catchment, intensity of rainfall and its duration and soil type.

SOLUTION 4.12
The *NCEES Handbook version 2.0,* Section 6.5.6.4 National Weather Service IDF Curve Creation is used to solve this question as follows:

$$P = \frac{T_d + 20}{100} = \frac{10 \times 60 + 20}{100} = 6.2 \; in$$

$$i = \frac{60P}{T_d} = \frac{60 \times 6.2}{10 \times 60} = 0.62 \; in/hr$$

Correct Answer is (D)

SOLUTION 4.13
The *NCEES Handbook version 2.0,* Section 6.5.6.5 Distance Weighing:

$$P = \frac{\sum_{i=1}^{3}\left(\frac{P_i}{L_i^2}\right)}{\sum_{i=1}^{3}\left(\frac{1}{L_i^2}\right)} = \frac{\frac{3}{5^2} + \frac{1}{3.5^2} + \frac{7}{10^2}}{\frac{1}{5^2} + \frac{1}{3.5^2} + \frac{1}{10^2}} = 2.1 \; in$$

Correct Answer is (B)

SOLUTION 4.14
The *NCEES Handbook version 2.0* Section 6.5.1.1 General Probability has different scenarios, in which case Section 6.5.1.3 Risk or Annual Exceedance Probability (AEP) applies to this question:

$$P = 1 - \left(1 - \frac{1}{T}\right)^n$$
$$= 1 - \left(1 - \frac{1}{25}\right)^{10}$$
$$= 0.34$$

Correct Answer is (A)

SOLUTION 4.15

The *NCEES Handbook version 2.0*, Section 6.5.8.6 Infiltration/ Horton Model is applied to this question as follows:

$$f = (f_o - f_c)e^{-kt} + f_c$$

In the above, f_o is the initial infiltration capacity which equals to $2.6\ in/hr$ as observed on the curve. f_c is the saturated infiltration capacity which equals to $0.45\ in/hr$ as observed on the curve.

Based on this information, the above formula can be rewritten as follows:

$$f = (2.6 - 0.45)e^{-kt} + 0.45$$
$$= 2.15e^{-kt} + 0.45$$

In order to solve the above formula for k, any point can be selected on the infiltration curve and substituted in the formula as follows:

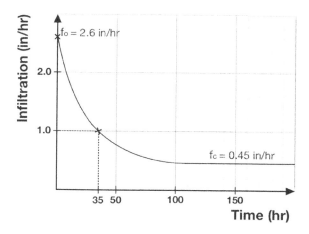

$$f = 2.15e^{-kt} + 0.45$$
$$1 = 2.15e^{-35k} + 0.45$$
$$e^{-35k} = 0.2558$$
$$k = \frac{\ln(0.2558)}{-35} = 0.039$$

The final *Horton* model/equation is presented as follows:

$$f = 2.15e^{-0.039t} + 0.45$$

You can take couple of points on the curve to check the validity of this model.

Correct Answer is (D)

SOLUTION 4.16

The *NCEES Handbook version 2.0*, Section 6.5.8.1 Surface Water System Hydrologic Budget is used to solve this question.

Start with converting all inflows and outflows into a consistent unit, in which case "ft" – and remember to use the study period in this case which is one year:

$$Q_{in} = \frac{10\frac{ft^3}{sec} \times \frac{(365 \times 24 \times 60 \times 60)sec}{yr}}{800\ acre \times \frac{43{,}560\ ft^2}{acre}} = 9.0\ ft$$

$$Q_{out} = \frac{7\frac{ft^3}{sec} \times \frac{(365 \times 24 \times 60 \times 60)sec}{yr}}{800\ acre \times \frac{43{,}560\ ft^2}{acre}} = 6.3\ ft$$

$$\Delta S_s = \frac{120\ acre.ft}{800\ acre} = 0.15\ ft$$

$$P = 9.5\ in \times \frac{ft}{12\ in} = 0.8\ ft$$

Apply the budget equation while removing all zero items:

$$P + Q_{in} - Q_{out} - E_s = \Delta S_s$$
$$0.8 + 9.0 - 6.3 - E_s = 0.15$$
$$E_s = 3.35\ ft\ (= 40.2\ in)$$

Correct Answer is (C)

SOLUTION 4.17

The *NCEES Handbook version 2.0*, Section 6.5.8.1 Surface Water System Hydrologic Budget is used to solve this question.

$$P + Q_{in} - Q_{out} + Q_g - E_s - T_s - I = \Delta S_s$$

Removing all zero items (Q_{in}, Q_g, T_s & I) and equating evaporation to 50% of the precipitation $E_s = 0.5P$, while the outflow Q_{out} is given as follows using a unit of depth over the study period of one month:

$$Q_{out} = \frac{5\frac{ft^3}{sec} \times \frac{(30 \times 24 \times 60 \times 60)sec}{month}}{500 \; acre \times \frac{43{,}560 \; ft^2}{acre}} = 0.6 \; ft$$

Apply the budget equation:

$$P + Q_{in} - Q_{out} + Q_g - E_s - T_s - I = \Delta S_s$$
$$P + 0 - 0.6 \; ft + 0 - 0.5P - 0 - 0 = 0$$
$$P - 0.6 \; ft - 0.5P = 0$$

$$\rightarrow P = 1.2 \; ft \; (= 14.4 \; in)$$

Correct Answer is (A)

SOLUTION 4.18

The *NCEES Handbook*, Section 6.6.3.1, Unconfined Aquifers/Dupuit's equation can be used to solve this question.

$$Q = \frac{\pi K (h_2^2 - h_1^2)}{\ln\left(\frac{r_2}{r_1}\right)}$$

Q is the flow rate in ft^3/sec, h_1 and h_2 are heights of the aquifer measured from its bottom at the perimeter of the well (i.e., $r_1 = \frac{12}{2} \; in = 0.5 \; ft$) and at the influence radius of $r_2 = 450 \; ft$ respectively.

Radius of influence defines the outer radius of the cone of depression, hence $h_2 = 100 \; ft$.

Check the below figure for more clarity (sketch not to scale):

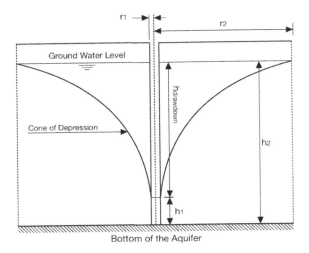

$$h_1 = \sqrt{\frac{\pi K h_2^2 - Q \times \ln\left(\frac{r_2}{r_1}\right)}{\pi K}}$$

$$= \sqrt{\frac{\pi \times 4\times 10^{-4}\frac{ft}{sec} \times (100 \; ft)^2 - 65\frac{gal}{min}\left(\frac{0.134 \; ft^3}{gal}\right)\left(\frac{1 \; min}{60 \; sec}\right) \times \ln\left(\frac{450 \; ft}{0.5 \; ft}\right)}{\pi \times 4 \times 10^{-4}\frac{ft}{Sec}}}$$

$$= 96 \; ft$$

$$h_{drawdown} = 100 \; ft - 96 \; ft = 4 \; ft$$

Correct Answer is (C)

SOLUTION 4.19

The *NCEES Handbook version 2.0*, Section 6.6.2.1 Darcy's Law is used to solve this question.

$$Q = -KA\frac{dh}{dx}$$

Where K is the permeability of soil given in the question as $0.0004 \; ft/sec$, A is the effective area per weep hole (see below sketch that shows an elevation view for the retaining wall and effective area per weep hole), and dh/dx is the hydraulic gradient, or the head over distance for water elevation given as $1:70$.

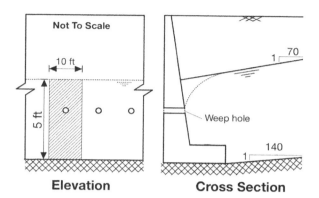

$$Q = -0.0004\ ft/sec \times (5 \times 10)\ ft^2 \times \frac{1}{70}$$

$$= -2.86 \times 10^{-4}\ cfs\ (= 24.7\ \frac{ft^3}{day})$$

Correct Answer is (D)

SOLUTION 4.20
Although this is a confined aquifer, with the limited information given in this question *Darcy's Law* can be used. *Darcy's* equation is found in the *NCEES Handbook version 2.0*, Section 6.6.2.1.

$$Q = -KA\frac{dh}{dx}$$

In this case, and given this is a confined aquifer, the area A in *Darcy's* equation corresponds to the area of the cylindrical boundary with radius $r = 130\ ft$ from the pumping well and thickness $b = 75\ ft$ – in fact, *Thiem* equation presented in Section 6.6.3.2 for confined aquifers is derived from *Darcy's* equation of Section 6.6.2.1.

Based on this, the above equation can be rewritten as follows:

$$Q = -(2\pi rb)K\frac{dh}{dx}$$

Where dh/dx is the hydraulic gradient given in the question as 0.03 at a distance $r = 130\ ft$ from the pumping well, and the value Kb represents transmissivity T.

$$Q = -(2\pi \times 130 \times 75)\ ft^2 \times 0.0005\frac{ft}{sec} \times 0.03$$

$$= -0.92\ cfs\ (\cong 413\ gpm)$$

Correct Answer is (A)

SOLUTION 4.21
Uniform flow by *Thiem* is used to solve this question, which is found in the *NCEES Handbook version 2.0*, Section 6.6.3.2.

$$Q = \frac{2\pi T(h_2 - h_1)}{\ln\left(\frac{r_2}{r_1}\right)}$$

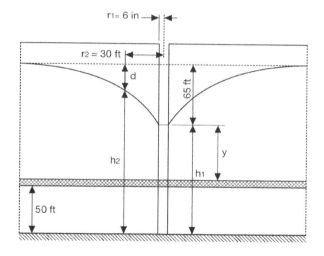

y is not important at this stage, shown here for demonstration purposes only:

$$h_1 = y + 50$$

$$h_2 = (65 + h_1) - d$$

$$h_2 - h_1 = 65 - d$$

$$Q = \frac{2\pi T(65-d)}{\ln\left(\frac{r_2}{r_1}\right)}$$

$$d = 65 - \frac{Q \times \ln\left(\frac{r_2}{r_1}\right)}{2\pi K b}$$

$$= 65 - \frac{75\ gpm \times \frac{cfs}{448.8\ gpm} \times \ln\left(\frac{30}{0.5}\right)}{2\pi \times 0.0004 \times 50}$$

$$= 59.6\ ft$$

Correct Answer is (D)

SOLUTION 4.22
The Multiple Aquifer Layers section is referred to in the *NCEES Handbook version 2.0,* page 420.

$$K = \sum_{i=1}^{4}\left[\frac{K_i b_i}{b}\right]$$

$$= \frac{0.001 \times 15 + 0.0005 \times 20 + 0.0035 \times 30 + 0.009 \times 45}{15 + 20 + 30 + 45}$$

$$= 0.0049\ ft/sec$$

Correct Answer is (B)

SOLUTION 4.23
The Seepage Velocity section, Section 6.6.2.3, is referred to in the *NCEES Handbook version 2.0*.

$$V_v = \frac{K}{\eta}\frac{\Delta h}{L}$$

Where η is the porosity of the aquifer and V_v is seepage velocity. Based on this, the formula can be rearranged as follows:

$$\eta = \frac{K}{V_v}\frac{\Delta h}{L}$$

$$V_v = \frac{1.5\ mile \times \frac{5,280\ ft}{mile}}{3\ yr \times \frac{365\ day}{yr}} = 7.2\ ft/day$$

$$\eta = \frac{85\ ft/day}{7.2\ ft/day} \times \frac{9\ ft}{(1.5 \times 5,280)ft}$$

$$= 0.013$$

Correct Answer is (A)

SOLUTION 4.24
To solve this question, we need to calculate the specific yield S_y, which represents the volume of water that drains from the aquifer by gravity per unit volume of the aquifer.

Specific yield S_y and specific retention S_r (the latter represents the water retained in the aquifer) both define porosity per the *NCEES Handbook version 2.0* page 418:

$$\eta = S_y + S_r$$

$$0.45 = S_y + 0.2$$

$$S_y = 0.25$$

In reference to the definition of specific yield:

$$S_y = \frac{V_{yield}}{V_{total}}$$

$$V_{total} = \frac{V_{yield}}{S_y}$$

$$= \frac{12\ ft \times 100\ acre}{0.25}$$

$$= 4,800\ ft.\ acre$$

$$Thickness = \frac{4,800\ ft.\ acre}{100\ acre} = 48\ ft$$

Correct Answer is (C)

V
ANALYSIS & DESIGN

Knowledge Areas Covered

SN Knowledge Area

4 Analysis and Design
  A. Mass balance
  B. Hydraulic loading
  C. Solids loading (e.g., sediment loading, sludge)
  D. Hydraulic flow measurement

9 Surface Water and Groundwater Quality
  A. Stream degradation and oxygen dynamics
  B. Total maximum daily load (TMDL) (e.g., nutrient contamination, DO, load allocation)
  C. Biological and chemical contaminants

10 Drinking Water Distribution and Treatment
  A. Drinking water distribution systems
  B. Drinking water treatment processes
  C. Present, short-term, and long-term demands
  D. Storage
  E. Sedimentation
  F. Coagulation and flocculation
  G. Membrane processes and media filtration
  H. Disinfection, including disinfection byproducts
  I. Hardness and softening
  J. Other treatment (e.g., ion exchange, carbon adsorption, ozone, UV, specific constituent removal)

11 Wastewater Collection and Treatment
  A. Wastewater collection systems (e.g., lift stations, sewer networks, infiltration, inflow, smoke testing, maintenance, odor control)
  B. Wastewater treatment systems
  C. Preliminary treatment
  D. Primary treatment
  E. Secondary treatment (e.g., physical, chemical, biological processes)
  F. Nutrient removal
  G. Solids treatment, handling, and disposal
  H. Disinfection
  I. Advanced treatment (e.g., advanced oxidation process, effluent filtration, adsorption, reclaimed water)

PART V
Analysis & Design

PROBLEM 5.1 *Influent Disinfection*

Solution 1 is inflowing into a 100,000 ft^3 tank at a rate of 30 cfs.

To treat this influent, it has been determined that *chlorine* inside the tank shall be maintained at a concentration level of no less than 3 mg/L.

In order to maintain this concentration, the tank is provided with a continuous flow of Solution 2 that has 750 mg/L of *chlorine*, while the tank maintains its volume with an outflow equivalent to flows of Solution 1 and 2 combined.

With a 0.5 mg/L per hour of *chlorine* depletion in this process (measured to the volume of the tank), the flow of Solution 2 into the tank in cfs should not be less than (*):

(A) 1.23

(B) 0.31

(C) 0.14

(D) 28

(*) Assume that the tank constituents are mixed well and that the constituents are at a continuous steady state.

PROBLEM 5.2 *Stormwater Drain Mixing*

A stormwater pipe with a flow of 5 cfs carries 1,500 mg/L concentration of deicing agent *sodium chloride* discharges into a storm water collection channel.

The water in the channel flows at a rate of 125 cfs and carries 4 mg/L of *sodium chloride* before the point of discharge (*).

Given the above information, the concentration of the deicing agent in the channel after the point of discharge is most nearly:

(A) 61.5 mg/L

(B) 1,440 mg/L

(C) 752 mg/l

(D) 98.6 mg/L

(*) Assume full mixing after the point of discharge.

PROBLEM 5.3 *Lake Pollutant Decay*

A 200×10^5 ft^3 lake is fed by a river flowing at a rate of 0.05 cfs and carrying a pollutant with a concentration of 25 mg/L (*).

Precipitation during the study year adds 500,000 ft^3 to the lake. Evaporation for the study year is estimated to reduce storage by 1,000,000 ft^3.

A tunnel was added to the river that takes away any excess water to generate a hydroelectric power plant.

Assuming the pollutant decays inside the lake at a rate of 0.15 per year, the expected concentration of this pollutant at the power plant during the study year is most nearly:

(A) 9.7 mg/L

(B) 13.1 mg/L

(C) 1.3 mg/L

(D) 21.4 mg/L

(*) Assume full mixing after the point of river discharge.

PROBLEM 5.4 *Water Treatment Plant Coagulation*

A 750×10^5 ft^3 water coagulation – flocculation treatment basin receives water at a flow rate of 20 cfs along with a 45 mg/L dose of *aluminum sulfate*. Suspended solids' concentration in this basin is 25 ppm, and the effluent outflows at a rate of 20 cfs and has its suspended solids concentration reduced to 9 mg/L due to coagulation (*).

Based on the above, the solid mass generated per day from this basin is most nearly:

(A) 6,582 lbm/day

(B) 110 lbm/day

(C) 3,885 lbm/day

(D) 1,727 lbm/day

(*) Assume full mixing of the coagulant in the basin.

PROBLEM 5.5 *Inline Equalization Tank Volume Calculation*

The below diagram represents an inflow mass diagram for a 1 MGD wastewater treatment plant.

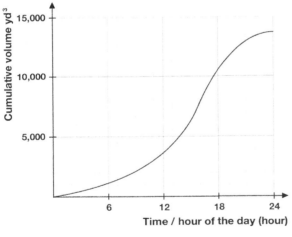

Inflow Mass Diagram

To prevent short term high volumes of incoming flows and to smoothen the flow for the treatment plant, it has been decided to install an inline equalization tank prior to the primary treatment works.

Based on the above diagram, the theoretical (*) volume of this tank is most nearly:

(A) 0.15 *million Gallon*

(B) 0.85 *million Gallon*

(C) 1.0 *million Gallon*

(D) 1.25 *million Gallon*

(*) Theoretical in this context means without providing extra allowances for aeration or any other equipment and without accounting for a safety factors.

PROBLEM 5.6 *Primary Settling Tank Hydraulic Loading*

A 100 ft diameter primary settling tank with an average depth of 10 ft has its Hydraulic Loading Rate/Surface Overflow Rate limited to 1,000 $Gallons/day/ft^2$ as recommended per Section 72.21 of the *Recommended Standards for Wastewater Facilities, 2014*.

The volumetric flow rate in MGD that can be treated in this settling tank is most nearly:

(A) 7.85

(B) 31.4

(C) 0.8

(D) 3.1

PROBLEM 5.7 *Offline Equalization Tank Volume Calculation*

The below table represents the inflow mass for a 0.5 MGD wastewater treatment plant.

In order to smoothen the flow for the treatment plant, and also to store some of the daily excess flow that exceeds the plant's capacity, an offline equalization tank is provided parallel to the primary treatment works.

Time	Flow
	ft^3/hr
12:00 AM - 06:00 AM	8,000
06:00 AM - 12:00 PM	39,000
12:00 PM - 06:00 PM	15,000
06:00 PM - 12:00 AM	22,000

Based on this, the depth of a circular $75\,ft$ diameter offline equalization tank should be:

(A) $7.5\,ft$

(B) $9.5\,ft$

(C) $12.2\,ft$

(D) $24.5\,ft$

PROBLEM 5.8 *Rapid Filter Loading Rate*
A rapid filter plant with five $20\,ft$ long by $20\,ft$ wide and $8\,ft$ deep rapid sand filters receive $25\,cfs$. The loading rate in gpm/ft^2 (*) is most nearly:

(A) $5.6\,gpm/ft^2$

(B) $28.6\,gpm/ft^2$

(C) $3.6\,gpm/ft^2$

(D) $0.7\,gpm/ft^2$

(*) gpm = Gallon Per Minute

PROBLEM 5.9 *Organic Loading*
The organic loading into a $50\,ft$ diameter, $10\,ft$ deep, primary settling tank that has an influent flow of $0.5\,MGD$ with a measured BOD concentration of $350\,mg/L$ is:

(A) $0.19\,lb/day$ BOD per ft^2

(B) $0.74\,lb/day$ BOD per ft^2

(C) $0.07\,lb/day$ BOD per ft^2

(D) $0.02\,lb/day$ BOD per ft^2

PROBLEM 5.10 *Increase Depth of a Flow Using of a Weir*
A sharp crested weir is to be installed into a rectangular channel to increase the depth of its flow from $1.5\,ft$ (dotted line) to $3.5\,ft$ (solid line).

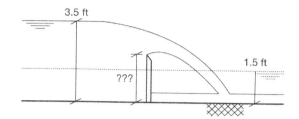

With a coefficient of discharge of $C = 0.57$, and a water velocity of $1\,ft/sec$ in the channel, the required height of the weir to achieve this increase in water depth is:

(A) $2\,ft$

(B) $1.9\,ft$

(C) $1.6\,ft$

(D) $4.3\,ft$

PROBLEM 5.11 *Parshall Flume Design*
A manufacturer designed a Parshall flume with a throat width of $W = 5\,ft$ along with the following flow data measurements:

Upstream head H_a (ft)	Flow Q (ft^3/sec)
0.25	4.88
0.5	12.1

The general flow formula for this flume is:

$$Q = CWH_a^{nW}$$

Based on the above flow measurements, values for C and n are as follows:

$C = $ ___

$n = $ ___

PROBLEM 5.12 *Ogee Weir Design*

The following WES (*) Ogee weir has a maximum flow of 150,000 cfs and a maximum total energy head over crest of 54 ft.

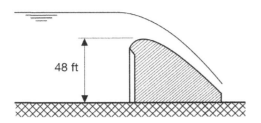

Assuming the ratio of the maximum head to design head is '1.24', the crest length with no piers installed should be:

(A) 91.5 ft

(B) 41.5 ft

(C) 48 ft

(D) 62.3 ft

(*) WES stands for Waterways Experiment Station.

PROBLEM 5.13 *Design of a Primary Settling Tank*

A primary settling tank is receiving waste activated sludge with the following data:

- Minimum hydraulic residence time = 2.0 hrs
- Size: circular with 35 ft diameter and 10 ft deep

Based on the above information, the design average flow for this tank in MGD is most nearly:

(A) 0.67

(B) 0.86

(C) 0.12

(D) 0.96

PROBLEM 5.14 *V-Notch Weir*

A Standard USBR (*) 90° V-notch weir is used in a channel, and it measures a flow rate of 75 cfs.

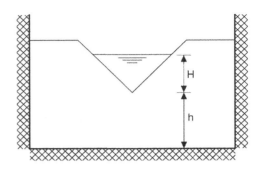

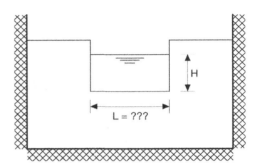

The length (L) of a USBR (*) sharp crested weir with end contractions that produces the same head at the same discharge is:

(A) 3.9 ft

(B) 3.7 ft

(C) 3.5 ft

(D) 4.7 ft

(*) The U.S. Bureau of Reclamation (USBR) standard coefficients of discharge for the above weir types are as follows:

Weir type	Coefficient of Discharge C
90° V-notch weir	2.49
Sharp crested weir with end contraction	3.33

PROBLEM 5.15 *Rapid Mixing*

A water treatment plant receives its a daily flow into a $3,000\ ft^3$ rapid mixer where it is introduced into a coagulating compound. A jar test determined that the velocity gradient for this mixer should be $950\ sec^-$ and that detention time should be $1.0\ minute$ when water is at $25°\ C$.

Given the above information, the estimated head loss due to friction generated from this operation is most nearly:

(A) $33.5\ ft$

(B) $1.6\ ft$

(C) $16.2\ ft$

(D) $3.4\ ft$

(✱) PROBLEM 5.16 *Flocculator Design*

The volume of the water treatment plant's first flocculator tank is $12\ ft \times 12\ ft \times 40\ ft$ as shown on the plan and sections below.

A jar test has determined that the velocity gradient for this mixer should be $50\ sec^-$ and the detention time should be $45\ minutes$ when water is at $32°\ C$ for flocculation to work without breaking down the flocs.

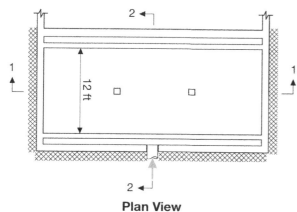

Plan View
Flocculation tank plan and section details in this question are for demonstration purposes only

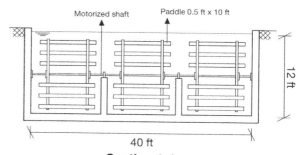

Section 1-1
The first tank with paddles showing

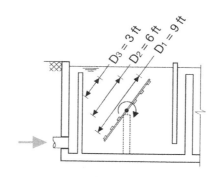

Section 2-2
Showing the first tank with paddles dimensions center to center

Assume that the speed of the blades relative to the water is 75% of the peripheral blade speed.

Based on the above information, and the dimensions given in the relevant cross sections, the rotation speed in *rpm* (rotation per minute) for those paddles should be:

(A) $1\ rpm$

(B) $2\ rpm$

(C) $3\ rpm$

(D) $4\ rpm$

PROBLEM 5.17 Adsorption using Powdered Activated Carbon PAC

The 0.5 MGD water treatment plant effluent contains 50 ug/L Trichloroethylene possibly due to an industrial leakage near the water intake, while the permitted limit for Trichloroethylene in drinking water is set as 5.0 ppb per WHO standards.

Using the following Freundlich adsorption isotherm coefficients for Trichloroethylene:

- $K_f = 28 \ (mg/g)(L/mg)^{1/n}$
- $1/n = 0.62$

The amount in $lb's$ of Powdered Activated Carbon (PAC) that shall be added daily to the water treatment works to bring the effluent into compliance with the WHO standards is:

(A) 100 lb/day

(B) 180 lb/day

(C) 330 lb/day

(D) 410 lb/day

PROBLEM 5.18 Required Air Flow for Air Stripping

It is desired to reduce the concentration of chloroform in a 2 MGD water treatment plant by 50% using a countercurrent packed tower (*) when temperature is $25°\ C$.

Given the Henry's law constant for chloroform at $25°\ C$ is 0.138, the required rate of air flow at the tower needed to achieve this stripping is most nearly:

(A) 5 cfs

(B) 15 cfs

(C) 40 cfs

(D) 80 cfs

(*) Assume that the optimal operating ratio is 7 times the minimum ratio for contaminant.

PROBLEM 5.19 Noncarbonate Hardness

The following table represents a water sample analysis for cations and anions measured in MEQ/L (milliequivalents per liter).

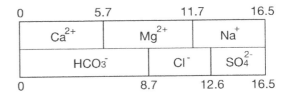

The noncarbonate hardness for this sample expressed as $mg/L\ CaCO_3$ is:

(A) 150

(B) 585

(C) 195

(D) 390

PROBLEM 5.20 Sedimentation Efficiency

A 20 ft deep rectangular sedimentation tank with a surface area of $25\ ft \times 75\ ft$ receives a flow rate of 4.5 MGD that contains the following particles' properties:

Average Settling velocity ft/hr	Number of particles per mL
3	1,300
6	1,800
9	2,000
12	1,100
Total	6,200

Given the above information, the sedimentation removal efficiency of those particles is most nearly:

(A) 20%

(B) 40%

(C) 55%

(D) 75%

PROBLEM 5.21 Water Distribution Reservoir

The below graph presents fluctuations of the hourly water demand (i.e., required flow) represented by the solid line.

The continuous flow provided by the treatment facility is represented by the dashed straight line (i.e., provided flow).

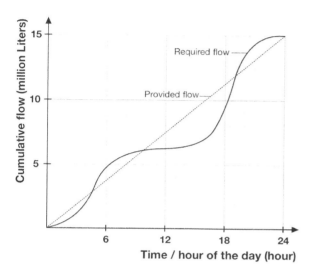

In order to balance (or equalize) storage and to provide a continuous flow for this fluctuating demand, ignoring fire storage, the following surface circular reservoir that accounts for 25% excess in volume as a safety factor is best used for this purpose:

(A) 95 ft dia, 18 ft deep

(B) 105 ft dia, 18 ft deep

(C) 110 ft dia, 20 ft deep

(D) 110 ft dia, 30 ft deep

PROBLEM 5.22 Population Growth

A city has a current population of 75,000 and an average water consumption of 5 MGD. The existing water treatment plant has a design capacity of 7.5 MGD. The population is expected to experience compounding growth until it reaches 165,000 in the next 20 $years$.

The number of years the plant can operate until it reaches its capacity is most nearly:

(A) 6 $years$

(B) 8 $years$

(C) 10 $years$

(D) 12 $years$

PROBLEM 5.23 Concentration Gradient

The concentration gradient for a filtration process with a solute influent concentration of 2,000 mg/L, an effluent concentration of 75 mg/L and a retentate concentration of 15 kg/m^3 is most nearly:

(A) 7.2 kg/m^3

(B) 15.0 kg/m^3

(C) 9.5 kg/m^3

(D) 8.6 kg/m^3

PROBLEM 5.24 Depth of a Filter Bed

A water treatment plant's filter bed has a loading area of 100 ft^2 and it consists of uniform (mono) sand. Water passes through this filter at a rate of 0.5 MGD. The filter's porosity is 0.45 and its sand's specific gravity is 2.65 and its shape factor is 0.85, and the diameter of particles is 0.0018 ft.

It is desired to maintain the head loss resulting from this filter at 3.5 ft when temperature is 70° F.

Based on the above information, the depth of this filter should be:

(A) 2 ft

(B) 3 ft

(C) 4 ft

(D) 5 ft

PROBLEM 5.25 *Disinfection*

The following statement(s) are true when it comes to disinfection using *chlorine*:

I. The concentration of *hypochlorous acid* and *hypochlorite* in pure water that receives *chlorine* as disinfectant is always similar. This is not the case however with water that contains impurities.
II. *Chlorine* residual in pure water will always equal to the *chlorine* dose.
III. Adding *chlorine* to surface water will result in a nonlinear relationship at the start of the process until it peaks at a certain concentration.
IV. Beyond the breakpoint of chlorination, relationship between *chlorine* dose and chlorine residual in groundwater is linear.
V. *Monochloramine* is predominantly formed in surface water while adding *chlorine* when all impurities have been oxidized.

(A) II + IV + V
(B) I + II
(C) IV + V
(D) II + III + V

PROBLEM 5.26 *Giardia Cysts 3-LOG Inactivation by Free Chlorine*

Residual *chlorine* is measured in a small groundwater system as 1.2 mg/L. To provide contact time, a 5,000 *gallon* baffled tank is provided with a baffling factor of '0.7'. The flow that passes through this tank is 150 *gpm* (gallon per minute).

With a water $pH = 7.0$, temperature of $20°\,C$, the actual contact time to the required contact time ($CT_{actual}/CT_{required}$) for 3-Log *giardia cysts* inactivation is most nearly:

(A) 0.5
(B) 1.2
(C) 2.3
(D) 3.4

PROBLEM 5.27 *Minimum Chlorine Residual*

When disinfection is accomplished using ultraviolet light (UV), *chlorine* is usually added to provide a minimum free *chlorine* residual in the water distribution system of:

(A) The use of UV excludes the need of other disinfectants in the system.
(B) 0.1 mg/L
(C) 0.2 mg/L
(D) 0.5 mg/L

PROBLEM 5.28 *Short Term Lagoon Sizing*

A 7 MGD water treatment plant that removes 150 mg/L hardness daily requires the installation of short-term lagoon(s).

The following lagoon(s) that can be used to store the resultant sludge for 2 ½ years:

☐ One lagoon with an area of 0.7 *acres* along with 5 ft usable depth.

☐ Two lagoons each with an area of 4.0 *acres* along with 5 ft usable depth.

☐ Three lagoons each with an area of 2.5 *acres* along with 4 ft usable depth.

☐ Three lagoons each with an area of 2.0 *acres* along with 5 ft usable depth.

☐ Four lagoons each with an area of 2.0 *acres* along with 5 ft usable depth.

PROBLEM 5.29 *Aeration Basin Dissolved Oxygen Concentration*

The below statements are true when it comes to the operation of an aeration basin's Dissolved Oxygen DO concentration in an activated sludge system:

I. Increasing the dissolved oxygen concentration in an aeration basin above 3 mg/L can improve treatment, but it is an expensive process.

II. Increasing the dissolved oxygen concentration in an aeration basis above 3 mg/L is not recommended as it can hinder the activation process.

III. Increasing the dissolved oxygen concentration in the aeration tanks can lead microorganisms to digest more contaminants.

IV. A minimum of 1 mg/L of dissolved oxygen should always be maintained to ensure proper treatment.

(A) I
(B) II + III
(C) I + III
(D) I + II + IV

PROBLEM 5.30 *Equalization Tank Aeration Requirements*

The concentration of dissolved oxygen in an equalization tank shall be maintained at a minimum of:

(A) 0 mg/L
(B) 1 mg/L
(C) 2 mg/L
(D) 3 mg/L

PROBLEM 5.31 *Wastewater Treatment Process Selection*

Match the described site location/size and characteristics/requirements on the left side with the proper secondary treatment process to the right side:

Location size and requirements	Secondary Treatment Process
Urban areas with compact solutions and the need for a high-quality effluent for reuse.	Activated Sludge Process (ASP)
Needed for efficient organic matter content removal, space is not an issue.	Trickling Filter
Needed mostly for organic matter reduction, decentralized system, and area with space constraints.	Membrane Bioreactor (MBR)
Needed for moderate organic load removal and a limited space for the system. It may need to be installed vertically.	Rotating Biological Contactor (RBC)

PROBLEM 5.32 Chlorination Injection into the RAS Line

Due to *filamentous* bacteria growth in an activated sludge operation, which does not settle well in the secondary clarifier like other microorganisms, it has been decided to inject *chlorine* into the Return Activated Sludge (RAS) line.

The RAS flow rate is 5 MGD and the concentration of its suspended solids' is 7,500 mg/L.

If a 0.5% of *chlorine* to the weight of the return suspended solids was required, the weight of *chlorine* to be injected in the RAS line in pounds per day is most nearly:

(A) 313 lb/day

(B) 1,564 lb/day

(C) 157 lb/day

(D) 188 lb/day

PROBLEM 5.33 Biological Denitrification by adding Methanol

A stage in a secondary treatment process has been added to remove *nitrate* and to convert it into *nitrogen* gas by adding *methanol* (CH_3OH).

The incoming flow into this stage is 0.5 MGD and it contains 25 mg/L nitrate, zero nitrite, and 8 mg/L dissolved oxygen.

The weight of *methanol* required to be injected into this process is most nearly:

(A) 423 lb/day

(B) 35 lb/day

(C) 290 lb/day

(D) 174 lb/day

PROBLEM 5.34 Dry Well Ventilation Requirements

A 7 ft × 6 ft × 8 ft pumping station's dry well is to receive continuous ventilation. The air required in CFM (Cubic Feet per Minute) to achieve this requirement is most nearly:

(A) 34 CFM

(B) 168 CFM

(C) 2,016 CFM

(D) 67 CFM

PROBLEM 5.35 Cascade Aeration System Design

The below is a cross section of a cascade step post-aeration system with a standard step height of 12 in.

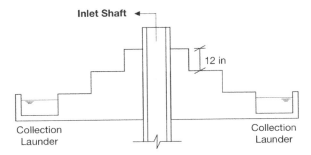

This aerator is located at sea level and the temperature could be as low as 59° F to a maximum of 86° F during different seasons. The sea level air solubility of oxygen at 86° F is 7.559 mg/L and at 59° F is 10.084 mg/L.

Assuming that the dissolved oxygen concentration in the influent to this aerator is 2.0 mg/L, while the desired concentration post cascade aeration is 6.0 mg/L, the number of steps for this aerator should be:

(A) 6

(B) 7

(C) 10

(D) 12

(⁂) PROBLEM 5.36 *Primary Anaerobic Sludge Digester*

A primary single-stage high-rate anaerobic sludge digestor serves a primary clarifier that receives wastewater flow of 5 MGD with 350 mg/L BOD and 250 mg/L TSS.

The primary clarifier achieves a 35% BOD reduction and 45% TSS removal.

Assuming the primary sludge has nearly a 94% moisture content and a specific gravity of 1.03 along with the following design parameters:

- Solid Residence Time $SRT = 20\ days$
- Process Efficiency $E = 70\%$
- Net mass of cell tissues produced $= 89.3\ lb/day$

The below design volumes can be determined as follows:

- Digester volume is: _____ ft^3
- Volume of *methane* (CH_4) produced is: _____ ft^3/day

(⁂) Normally you are requested to provide one value.

PROBLEM 5.37 *Primary Settling Tank Scour Velocity*

The peak flow rate for the below two primary settling tanks is 30 MGD.

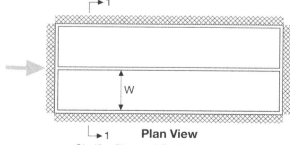

Plan View
Clarifier Plan and Section details in this question are for demonstration purposes only

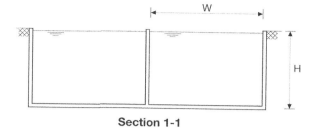

Section 1-1

The below single tank cross-section(s) ($W \times H$) shall satisfy the needed scour velocity (*) with a factor of safety of '2' (select all that applies):

(1) $30\ ft \times 8\ ft$

(2) $24\ ft \times 10\ ft$

(3) $20\ ft \times 10\ ft$

(4) $20\ ft \times 12\ ft$

(5) $25\ ft \times 8\ ft$

(*) Use the following parameters for the settled particles:

- $\beta = 0.05$ (constant for the type of material being scoured)
- Specific gravity of particles $= 1.25$
- Diameter of particles $= 3.28 \times 10^{-4}\ in$
- *Darcy-Weisbach* friction factor $= 0.025$

PROBLEM 5.38 *Trickling Filter BOD Removal*

A 60 ft diameter and 12 ft deep trickling filter with an influent flow rate of 0.2 MGD and a BOD concentration of 250 mg/L receives recycled flow at a rate of 1 cfs.

Based on the above information, the effluent's BOD concentration when temperature is $20^\circ\ C$ is:

(A) 28 mg/L

(B) 10 mg/L

(C) 23 mg/L

(D) 35 mg/L

PROBLEM 5.39 *Activated Sludge MLSS Thickening*

In a DAF system, the sludge mixed liquor flow rate that requires thickening is 0.5 MGD out of which 0.5% is solids.

In order to achieve an optimum air to solids ratio of '0.008' (typical) when the temperature is $30^0\ C$, the gauge pressure in kPa should read:

(A) $299.9\ kPa$

(B) $599.8\ kPa$

(C) $498.6\ kPa$

(D) $101.35\ kPa$

📖 PROBLEM 5.40 *Biological Denitrification /Biological Phosphorus Removal*

The following set of processes or zones represent a proper biological denitrification (*) process starting from left to right:

Option (A)

Anaerobic → Aerobic → Anaerobic → Aerobic

Option (B)

Anoxic → Aerobic → Anaerobic → Aerobic

Option (C)

Anoxic → Aerobic → Anoxic → Aerobic

Option (D)

Anaerobic → Anoxic → Aerobic → Anoxic

(*) Check throughout the solution to read more about the denitrification process and the biological phosphorus removal as well toward the end of the solution.

PROBLEM 5.41 *Food to Microorganism Ratio*

The following sample and flow characteristics where evaluated based on a flow sample collected from a treatment plant:

- Flow $Q = 3.0\ MGD$
- Tank volume $V = 200 \times 10^3\ ft^3$
- BOD Concentration $= 350\ mg/L$
- Sample volume $= 300\ mL$
- Weight of filtered sample after drying $= 633.2\ mg$
- Filter weight after baking $= 215.1\ mg$

Based on the above information, the F/M ratio for this treatment plant is most nearly:

(A) 0.5

(B) 0.6

(C) 0.7

(D) 0.8

PROBLEM 5.42 *Returned Activated Sludge Flow (RAS)*

The flow into an activated sludge treatment plant is 5 MGD. The estimated RAS Total TSS concentration is $6,500\ mg/L$ and MLSS concentration is $3,500\ mg/L$.

Based on the above information, the flow in the RAS line should be _____ MGD.

PROBLEM 5.43 *Dechlorination*

A treatment plant discharges its effluent into a river at a flow rate of 10 MGD. The effluent has a *chlorine* residual of $1.75\ mg/L$. The concerned environment agency requires the Cl_2 concentration to be $< 0.5\ mg/L$ prior to discharge. The weight of *sodium thiosulfate* $Na_2S_2O_3$ as a solution that should be mixed with the effluent to comply with this regulation is _____ lb/day.

PROBLEM 5.44 *Pond Water Quality*

The following are effective strategy(s) that can be used to safely treat large ponds for odor:

I. Aeration
II. Chemical treatment
III. Removing algae and other debris
IV. Bacterial treatment

(A) I

(B) II

(C) I + IV

(D) II + III

PROBLEM 5.45 *Stream Degradation caused by an Outfall*

A treatment plant discharges its effluent into a 7.5 ft deep river that runs at an average velocity of 2 ft/sec. The ultimate BOD just below this outfall is measured as 35 mg/L and the dissolved oxygen is 3.5 mg/L at the mixing zone when temperature is 20° C.

Based on the above information (*), the critical distance from the outfall to the point of maximum oxygen deficit is most nearly:

(A) 28 mile

(B) 1 mile

(C) 20 mile

(D) 39 mile

(*) Use *Churchill et al.* method to determine the reaeration rate coefficient. Also, use a deoxygenation coefficient with base e of $k_d = 0.6 \ day^{-1}$.

PROBLEM 5.46 *Oxygen Deficit caused by an Outfall*

A treatment plant discharges its effluent at a rate of 45 cfs into a river. The outfall's BOD is 20 mg/L and dissolved oxygen of 2.2 mg/L. The river flows at a rate of 175 cfs and has a BOD just upstream the outfall of 5.5 mg/L and dissolved oxygen of 6.5 mg/L.

Based on the above information, the initial deficit just after the outfall is most nearly:

(A) 0.9 mg/L

(B) 4.3 mg/L

(C) 5.6 mg/L

(D) 12.1 mg/L

PROBLEM 5.47 *Maximum Specific Growth Rate for Bacillus Subtilis*

A fermentation experiment for growing microorganism *Bacillus subtilis* on a *glucose* substrate resulted with the following data:

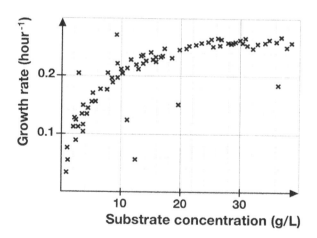

Based on this data capture, *Monod's* saturation constant is most nearly:

(A) 0.13 g/L

(B) 0.26 g/L

(C) 3.3 g/L

(D) 21 g/L

PROBLEM 5.48 *Drinking Water Equivalent Level (DWEL)*

During a course of a study for a certain carcinogenic chemical which was conducted on rats, the dose at which no effects are observed was identified as $0.3\ mg/kg$. As this study was conducted on rats, a 100 folds uncertainty factor is considered.

The drinking water equivalent level (DWEL) for this chemical for a lifetime consumption of 2.3 Liter per day for a 67 kg female is most nearly:

(A) $0.03\ mg/L$

(B) $0.003\ mg/L$

(C) $0.087\ mg/L$

(D) $0.0087\ mg/L$

PROBLEM 5.49 *Probability of Risk from Carcinogenic Contaminant*

A carcinogenic substance was discovered in drinking water which was consumed by a 75 kg adult at a rate of $2\ L/day$. The concentration of this substance was $0.02\ mg/L$ and it was consumed over the course of 40 years.

The cancer slope factor for this contaminant is $0.35\ kg.day/mg$.

The risk of cancer for this individual assuming a lifetime of 75 years is:

(A) 9.94×10^{-5}

(B) 9.94×10^{-6}

(C) 2.8×10^{-4}

(D) 2.8×10^{-5}

PROBLEM 5.50 *Total Maximum Daily Loads TMDLs*

The following statement(s) are true when it comes to TMDLs:

I. The main purpose of a TMDL is to understand and assess the severity of pollutants that are entering into a waterbody and warn the public.

II. The objective of a TMDL is to increase the loading capacity of waterbodies.

III. A discharge pipe is an example of a point source of a TMDL study.

IV. TMDL is a complex study that involves conducting mass balances, some complex modelling, and calculations, and most importantly, public involvement is key to such studies.

(A) I

(B) II

(C) III + IV

(D) II + III

SOLUTION 5.1

The principal of conservation of mass, in which case the mass balance equation, is applied to this solution.

The mass balance equation of the *NCEES Handbook* Section 6.7.1, or any of its derivatives, can be referred to. See below:

$$\frac{dM}{dt} = \frac{dM_{in}}{dt} + \frac{dM_{out}}{dt} \mp r$$

Where M represents the mass in this case, and while the inflow equals to the outflow, the change of the tank constituents' mass → $\frac{dM}{dt} = 0$. And r is the reaction rate, in which case it represents depletion, and it should be multiplied by the tank volume as referred to in the question.

Based on this, the above equation can be rewritten as follows:

$$0 = \frac{dM_{in}}{dt} - \frac{dM_{out}}{dt} - r$$

In reference to the below sketch, the above equation can be rewritten as shown below where Q_i is flow of Solution i, and $C_{cl,i}$ and $C_{cl,tank}$ are *chlorine* concentrations for Solution i and the tank respectively.

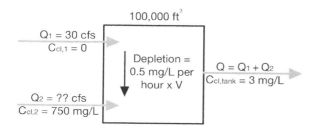

$$0 = (Q_1 C_{cl,1} + Q_2 C_{cl,2}) - (Q_1 + Q_2) C_{cl,tank} - V \times r$$

$$Q_2 = \frac{Q_1 C_{cl,tank} + V \times r}{C_{cl,2} - C_{cl,tank}}$$

$$= \frac{30\ cfs \times 3\ mg/L + 10^5 ft^3 \times \frac{0.5\ mg/L}{3,600\ sec}}{750\ mg/L - 3\ mg/L}$$

$$= 0.14\ cfs$$

Correct Answer is (C)

SOLUTION 5.2

The principal of conservation of mass, and the mass balance equation, is used in this solution.

The mass balance equation of the *NCEES Handbook* is referred to. A derivative of it is used per the below assuming no reaction with the *sodium chloride* takes please, which sets r to zero. Also, given full mixing as stated in the question (i.e., a steady flow state), this makes the change in volume $\left(\frac{dM}{dt}\right) = 0$.

Based on the above, the following equation can be established:

$$\sum Q_{in} C_{in} = \sum Q_{out} C_{out}$$

$$Q_p C_p + Q_c C_c = (Q_p + Q_c) C_{mixed}$$

$$C_{mixed} = \frac{Q_p C_p + Q_c C_c}{Q_p + Q_c}$$

$$= \frac{5\ cfs \times 1,500\ mg/L + 125\ cfs \times 4\ mg/L}{5\ cfs + 125\ cfs}$$

$$= 61.5\ mg/L$$

Correct Answer is (A)

SOLUTION 5.3

The principal of conservation of mass, and the mass balance equation, is used in this solution.

The rate of flow, the rate of discharge, and the rate of pollutant decay, in this question are all set to 'per year'. With this, start with determining the missing outflow rate which goes to the tunnel by balancing the given volumes as follows – taking units into account:

$$\sum Q_{in} = \sum Q_{out}$$

$$Q_{river} + Q_{rain} = Q_{evap} + Q_{tunnel}$$

$$\left[0.05 \frac{ft^3}{sec} \times 31.536 \times 10^6 \frac{sec}{yr} + 5 \times 10^5\right] ft^3/yr$$
$$= [1 \times 10^6 + Q_{tunnel}]\, ft^3/yr$$

$$Q_{tunnel} = 1{,}076{,}800\, ft^3/yr$$

Apply the mass balance equation of the *NCEES Handbook*, Section 6.7.1, or any of its derivatives, with $\left(\frac{dM}{dt} = 0\right)$ as follows:

$$\frac{dM}{dt} = \frac{dM_{in}}{dt} + \frac{dM_{out}}{dt} \mp r$$

$$0 = \frac{dM_{in}}{dt} - \frac{dM_{out}}{dt} - r$$

$$0 = \sum Q_{in} C_{in} - \sum Q_{out} C_{out} - V(kC_{out})$$

Where k is the decay rate per year and is multiplied by the overall concentration rate of the lake and the lakes' total volume. Also, the outflow pollutant concentration is the same as the lake's pollutant concentration. See below:

$$0 = (Q_{river} C_{river} + Q_{rain} C_{rain})$$
$$- (Q_{evap} C_{evap} + Q_{tunnel} C_{out})$$
$$- V(kC_{out})$$

In which case $C_{rain} = C_{evap} = 0$

$$0 = Q_{river} C_{river} - Q_{tunnel} C_{out} - VkC_{out}$$

$$C_{out} = \frac{Q_{river} C_{river}}{Q_{tunnel} + Vk}$$

$$= \frac{(0.05 \times 31.536 \times 10^6)\, ft^3/yr \times 25\, mg/L}{1{,}076{,}800\, ft^3/yr + 200 \times 10^5\, ft^3 \times 0.15/yr}$$

$$= 9.7\, mg/L$$

Correct Answer is (A)

SOLUTION 5.4

Apply the mass balance equation of the *NCEES Handbook* Section 6.7.1 as follows:

$$\frac{dM}{dt} = \frac{dM_{in}}{dt} + \frac{dM_{out}}{dt} \mp r$$

Given this is a steady state $\rightarrow \left(\frac{dM}{dt} = 0\right)$ with no reaction $\rightarrow (r = 0)$:

$$\frac{dM_{in}}{dt} = \frac{dM_{out}}{dt}$$

$$\sum Q_{in} C_{in} = \sum Q_{out} C_{out} + \frac{dM_{sludge}}{dt}$$

Also, it is important to remember that:
$$1\, ppm = 1\, mg/L$$

$$20\, cfs \times \left(45\, \frac{mg}{L} + 25\, \frac{mg}{L}\right)$$
$$= 20\, cfs \times 9\, \frac{mg}{L} + \frac{dM_{sludge}}{dt}$$

$$\frac{dM_{sludge}}{dt} = 20 \frac{ft^3}{sec} (45 + 25 - 9) \frac{mg}{L}$$

$$= 1{,}220 \frac{ft^3 \cdot mg}{sec \cdot L} \times \left(\frac{28.32\ L}{ft^3}\right) \times$$

$$\left(\frac{86{,}400\ sec}{day}\right) \times \left(\frac{2.205 \times 10^{-6}\ lbm}{mg}\right)$$

$$= 6{,}582.3\ lbm/day$$

Correct Answer is (A)

SOLUTION 5.5

In an inline equalization basin, the flow travels through the basin and directly into the treatment works. While in a sideline or offline equalization tank, only the flow above a predefined flow limit is diverted into the tank while pumping requirements are reduced.

The question asks to design an inline equalization tank, hence, the volume generated from the given inflow diagram will be used in full without deducting from it the daily limit of the treatment works.

To calculate the required tank volume, the following steps are performed:

1- The origin and the final points on the mass curve are connected with a dotted line. This line represents the average daily flow.

2- Two tangents to the two peaks of the mass curve are constructed, those should be parallel to the average flow line of step 1 – those two lines are represented with line 1 and line 2 on the below graph.

3- The required volume is the vertical distance between the two constructed lines of step 2, which has been determined on the graph as 4,200 cubic yards (0.85 million Gallon).

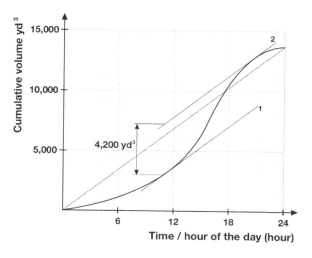

Usually, 25% is added to this volume as a safety factor. Since the question did not ask for this, the theoretical volume for this tank is:

4,200 *cubic yards* (0.85 *million Gallon*)

Correct Answer is (B)

SOLUTION 5.6

This problem can be solved by calculating the surface area of the settling tank and multiplying it by the given overflow rate as follows:

$$= 0.001 \frac{\frac{Million\ Gallons}{day}}{ft^2} \times \pi \times (50\ ft)^2$$

$$= 7.85\ MGD$$

Correct Answer is (A)

SOLUTION 5.7

Similar to the solution of question 5.5, the intent here is to construct a cumulative inflow mass diagram by way of adding more context into the suggested table as follows:

I	II	III	IV	V	VI
Period	Flow $\times 10^3$ ft^3/hr	Flow per period $\times 10^3$ ft^3	Average flow per period $\times 10^3$ ft^3	III – IV	Cumulative for column (V)
1st 6 hrs	8	48	126	−78	−78
2nd 6 hrs	39	234	126	108	30
3rd 6 hrs	15	90	126	−36	−6
4th 6 hrs	22	132	126	6	0
Sum		504			

The first column is adjusted to target a period of six hours. There are four such periods.

Column II is given in the question.

Column III determines the total flow for each of those 6-hour periods by multiplying the value in column II by 6 hours.

Column IV calculates the average flow rate for each 6-hour period. This is done to model what flow it takes for the treatment plant to operate smoothly without the sudden changes in flow rate. It also represents the dotted line of solution 5.5. This is calculated as follows:

$$\frac{\sum 24 \, hrs \, flow}{24 \, hrs} \times period = \frac{504 \times 10^3}{24} \times 6$$

$$= 126 \times 10^3 \, ft^3$$

Column V represents the variance to the flow from the desired average, and column VI is the cumulative of that, which is a representation of the inflow mass curve.

Based on the above, and in order to smoothen the flow for the treatment plant, an equalization tank's volume that equals to the difference between the lowest and the greatest value on the cumulative curve of column VI shall be provided as follows:

$$V = [30 - (78)] \times 10^3 = 108 \times 10^3 \, ft^3$$

$$= 0.81 \, Million \, Gal$$

When the plant capacity is $0.5 \, MGD$, a tank volume of $0.31 \, Million \, Gallon$ shall be provided offline to the operation as follows:

$$V = 0.81 - 0.5 = 0.31 \, MG \, (= 41{,}540 \, ft^3)$$

The depth of a circular $75 \, ft$ equalization tank is therefore calculated as follows:

$$d = \frac{41{,}540 \, ft^3}{\pi \times (37.5 \, ft)^2} = 9.4 \, ft$$

Normally, a safety factor should be applied to this volume, but it is not requested in this question.

Correct Answer is (B)

SOLUTION 5.8

Loading rate equals to the flow rate divided by (the top or the loading) surface areas for all filters – depth is not used in this case:

$$LR = \frac{25 \, \frac{ft^3}{sec} \times \frac{7.463 \, \frac{gal}{ft^3}}{\frac{1 \, min}{60 \, sec}}}{5 \times (20 \, ft \times 20 \, ft)} = 5.6 \, gpm/ft^2$$

Correct Answer is (A)

SOLUTION 5.9

Use the *NCEES Handbook* Section 6.8.5.3 Activated Sludge Treatment, Solids Loading equation in page 447 of the handbook's version 2.0 as follows:

$$SL = 8.34 \, QX$$

$$= 8.34 \times 0.5 \, MGD \times 350 \, mg/L$$

$$= 1{,}459.5 \, lb/day$$

To determine the requested loading, this value shall be divided by the area of the tank as follows:

$$= \frac{1{,}459.5 \, lb/day}{\pi \times 25^2 \, ft^2}$$

$$= 0.74 \frac{lb}{day} \, BOD \, per \, ft^2$$

Correct Answer is (B)

SOLUTION 5.10

Use the *NCEES Handbook* Section 6.2.5.4 Rectangular Weirs, the Sharp-Crested Weirs equation is used as follows:

$$Q = CLH^{3/2}$$

Where L is the width of the channel (i.e., the length of the weir), which is not given in this question. However, flow Q can be calculated using L and the initial water depth ($1.5 \, ft$), along with the given velocity. When Q is equated to the above equation, L becomes a common factor that can be eliminated.

$$Q = V \times A$$

$$= 1 \, ft/sec \times (1.5 \, ft \times L)$$

$$= (1.5 \, L) \, ft^3/sec$$

Substitute this value into the initial equation:

$$Q = CLH^{3/2}$$

$$1.5 \, L = 0.57 L H^{3/2}$$

$$H^{3/2} = 2.63$$

$$H = 2.63^{2/3} = 1.9 \, ft$$

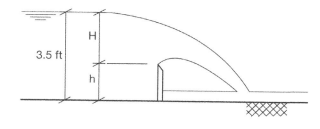

$$\rightarrow h = 3.5 - 1.9 = 1.6 \, ft$$

Correct Answer is (C)

SOLUTION 5.11

This question is solved by substituting the flow information in the general flume equation. Given the equation has two unknowns, two iterations will be required.

Using $H_a = 0.25$:

$$Q = CWH_a^{nW}$$

$$4.88 = 5C(0.25)^{5n}$$

Using $H_a = 0.5$:

$$Q = CWH_a^{nW}$$

$$12.1 = 5C(0.5)^{5n}$$

Divide the two equations by each other to remove C:

$$\frac{12.1}{4.88} = \frac{5C(0.5)^{5n}}{5C(0.25)^{5n}}$$

$$2.48 = (2)^{5n}$$

Using the mathematical equations and identifiers from the *NCEES Handbook's* first chapter:

$$5n = \frac{\ln(2.48)}{\ln(2)}$$

$$n = 0.262$$

Substituting this value in any of the above equations leads to C as follows:

$$C = \frac{4.88}{5(0.25)^{5 \times 0.262}} = 6$$

The final equation will therefore look like:

$$Q = 6WH_a^{0.262W}$$

SOLUTION 5.12

An Ogee weir spillway is used to transfer large flood discharges safely from a reservoir or a dam into a downstream flow with a significant elevation change. The design of such a spillway requires several iterations and a physical model may need to be built as well.

As for this question, along with the given assumptions, the *NCEES Handbook* Section 6.2.7 can be referred to.

The following information were given in the question – also refer to the Ogee figure in the handbook:

$Q = 150,000 \ cfs$

$P = 48 \ ft$

$H_e = H_d + H_a = 54 \ ft$

$\left(\frac{H_d + H_a}{H_d}\right) = 1.24$

$\rightarrow H_d = 54/1.24 = 43.55$

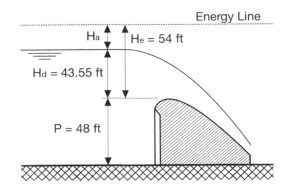

Based on the above, the following two values can be determined for later use:

$$\frac{P}{H_d} = \frac{48}{43.55} = 1.1$$

$$\frac{H_e}{H_d} = \frac{54}{43.55} = 1.24 \ (Given)$$

The above two values can be used in the graph presented in the *NCEES Handbook* version 2.0 page 326 to determine the coefficient of discharge C as follows:

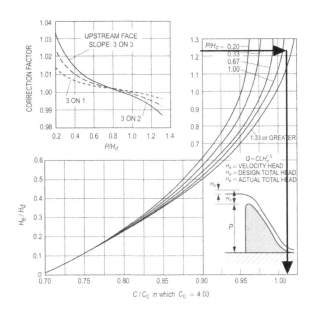

$$\frac{C}{C_o} = \frac{C}{4.03} = 1.025$$

$$\rightarrow C = 4.13$$

To calculate length of the crest L, reference is made to the flow equation as follows:

$$Q = CLH_e^{3/2}$$

$$\rightarrow L = \frac{Q}{CH_e^{3/2}} = \frac{150{,}000\ cfs}{4.13 \times (54)^{3/2}} = 91.5\ ft$$

Correct Answer is (A)

SOLUTION 5.13
Reference is made to Section 72.21 of the *Recommended Standards for Wastewater Facilities, 2014* where the surface overflow rate for <u>design average flows</u> for tanks <u>receiving waste activated sludge</u> should not exceed 700 gpd/ft^2.

The surface area for the given tanks is:

$$A = \pi \times 17.5^2 = 962.1\ ft^2$$

Surface loading (see *NCEES Handbook* Section 6.8.5.2):

$$v_o = Q/A$$

$$700\ gpd/ft^2 \geq Q/A$$

$$Q \leq v_o \times A$$

$$\leq 700\ gpd/ft^2 \times 962.1\ ft^2$$

$$\leq 673{,}470\ gallon\ per\ day\ (= 0.67\ MGD)$$

A check on the hydraulic residence time is also required where $\theta \geq 2.0\ hrs$:

$$\theta = V/Q$$

$$2.0\ hrs \leq V/Q$$

$$Q \leq V/2.0\ hrs$$

$$\leq \frac{(962.1 \times 10)\ ft^3 \times 7.463\ \frac{gal}{ft^3}}{2.0\ hrs \times \frac{day}{24\ hrs}}$$

$$\leq 860{,}722.7\ gallon\ per\ day\ (0.86\ MGD)$$

The minimum value shall be selected in this case which is $0.67\ MGD$.

Correct Answer is (A)

SOLUTION 5.14
The solution of this question requires a straight application for the two weirs equations found in the *NCEES Handbook* Section 6.2.5.4 as follows:

Determine H for the 90° V-notch weir:

$$Q = CH^{5/2}$$

$$\rightarrow H = \left(\frac{Q}{C}\right)^{2/5}$$

$$= \left(\frac{75}{2.49}\right)^{2/5}$$

$$= 3.9\ ft$$

Determine L for the sharp crested weir:

$$Q = C[L - n(0.1H)]H^{3/2}$$

$$\rightarrow L = \frac{Q}{CH^{3/2}} + n(0.1H)$$

$$= \frac{75}{3.33 \times 3.9^{3/2}} + 2 \times (0.1 \times 3.9)$$

$$= 3.7\ ft$$

Correct Answer is (B)

SOLUTION 5.15
Rapid mixing with a coagulating compound is normally the first process in any water treatment system, hence why the velocity gradient is quite high along with the possible generated head losses.

The velocity gradient equation of the *NCEES Handbook* Section 6.9.2.2 is used for this purpose as follows:

$$G = \sqrt{\frac{\gamma H_L}{t\,u}}$$

$$\rightarrow H_L = \frac{G^2\,t\,u}{\gamma}$$

Where u is the water's dynamic viscosity taken (interpolated) from Section 6.2.1.6 of the handbook as $1.86 \times 10^{-5}\ lbf.sec/ft^2$ for water at temperature of $25°\,C\,(= 77°\,F)$.

t is detention time given as $1.0\ minute$ and γ is the specific weight for water which is $62.23\ lb/ft^3$ using the same table of Section 6.2.1.6.

$$H_L = \frac{\left(950\,\frac{1}{sec}\right)^2 \times (1.0 \times 60)\ sec \times 1.86 \times 10^{-5}\,\frac{lbf.sec}{ft^2}}{62.23\,\frac{lbf}{ft^3}}$$

$$= 16.2\ ft\ (*)$$

Correct Answer is (C)

(*) The power required can also be determined using the same information and the same section of the handbook as follows:

$$G = \sqrt{\frac{P}{u\,V}}$$

$$\rightarrow P = G^2\,u\,V$$

$$= \left(\frac{950}{sec}\right)^2 \times 1.86 \times 10^{-5}\,\frac{lbf.sec}{ft^2} \times 3{,}000\ ft^3$$

$$= 50{,}360\ ft.lbf/sec$$

(✱) **SOLUTION 5.16**
The power to fluid is initially calculated using equations of the *NCEES Handbook* Section 6.9.2.2 as follows:

$$G = \sqrt{\frac{P}{u\,V}}$$

$$\rightarrow P = G^2\,u\,V$$

Where u is the water's dynamic viscosity taken from Section 6.2.1.6 of the handbook as $1.595 \times 10^{-5}\ lbf.sec/ft^2$ for water at temperature of $32°\,C\,(\cong 90°\,F)$.

$$P = G^2\,u\,V$$

$$= \left(\frac{50}{sec}\right)^2 \times 1.595 \times 10^{-5}\,\frac{lbf.sec}{ft^2} \times (12 \times 12 \times 40)\ ft^3$$

$$= 229.7\ ft.lbf/sec$$

The above power can be substituted in the following equation of the same section of the handbook to derive the blades' speed, hence the rotational revolution speed. Each blade however should be analyzed alone, and then summed up to the above power. See below:

$$P = \frac{C_D\,A_P\,\rho_f\,v_r^3}{2}$$

C_d is the drag coefficient which can be taken as '1.8' for blades with aspect ratio of '20' – i.e., the case of this question.

A_P is the area of blades. Since each blade has a different speed and will generate a different power, A_P is grouped into three blades based on their circumferential location:

1- The outer most blades located at diameter $D_1 = 9\ ft$ and area A_{P1}:

$$A_{P1} = 0.5\ ft \times 10\ ft \times 2\ blades \times 3\ mixers$$

$$= 30\ ft^2$$

2- The middle blade at $D_2 = 6\ ft$ and $A_{P2} = 30\ ft^2$

3- The inner most blade at $D_3 = 3\,ft$ and $A_{P3} = 30\,ft^2$

v_r is the relative speed which equals to the slip coefficient (given as $'0.75'$) multiplied by the velocity of each paddle v_p.

The paddles' velocity v_p is calculated as follows where N is the desired rotational speed measured in *revolution per sec*:

$$v_p = 2\pi r N = \pi D N$$

Based on the above information, the power equation can be rewritten as follows:

$$P = P_1 + P_2 + P_3$$

$$P_1 = \frac{C_D\, A_{P1}\, \rho_f\, v_{r1}^3}{2}$$

$$= \frac{C_D \times A_{P1} \times \frac{\rho_w}{g_c} \times (0.75 \times \pi \times D_1 \times N)^3}{2}$$

$$= \frac{1.8 \times 30\,ft^2 \times \frac{62.4\,\frac{lbm}{ft^3}}{32.174\,\frac{lbm.ft}{lbf.sec^2}} (0.75 \times \pi \times 9\,ft \times N)^3}{2}$$

$$= 499{,}349.1\,(N)^3\ ft.lbf/sec$$

Using the same logic, P_2 and P_3 can be calculated while changing the diameter to D_2 and D_3:

$$P_2 = 147{,}955.3\,N^3$$

$$P_3 = 18{,}494.4\,N^3$$

$$229.7 = 499{,}349.1\,N^3 + 147{,}955.3\,N^3 + 18{,}494.4\,N^3$$

$$\rightarrow N = 0.07\,rev/sec = 4.2\,rpm$$

Correct Answer is (D)

📖 SOLUTION 5.17

Reference is made to the *NCEES Handbook* Section 6.9.3.1 Freundlich Isotherm and the following equation:

$$K_f\, C_e^{1/n} = \frac{(C_o - C_e)V}{w}$$

Where V is volume in *Liters*, w is the PAC weight in *grams*, C_o and C_e are concentrations before and after the adsorption by PAC has occurred in mg/L.

Rearrange the equation to solve for weight/volume ratio as follows:

$$\frac{w}{V} = \frac{(C_o - C_e)}{K_f\, C_e^{1/n}}$$

$$= \frac{(50 \times 10^{-3} - 5 \times 10^{-3})\,mg/L}{28\left(\frac{mg}{g}\right)\left(\frac{L}{mg}\right)^{0.62} \times \left(5 \times 10^{-3}\,\frac{mg}{L}\right)^{0.62}}$$

$$= 0.043\,g/L$$

Multiply the above ratio by the effluent's daily volume after making the necessary unit conversions as follows:

$$= 0.043\,\frac{g}{L} \times \frac{2.205 \times 10^{-3}\,\frac{lb}{g}}{\frac{1}{3.785}\,\frac{gal}{L}} \times 0.5 \times 10^6\,\frac{gal}{day}$$

$$= 179.4\,lb/day$$

Correct Answer is (B)

📖 It is worth noting that a fixed-bed contactor using Granular Activated Carbon (GAC) is generally more advantageous. It achieves lower carbon usage per volume treated compared to the addition of Powdered Activated Carbon (PAC). However, the primary advantage of adding PAC in this specific problem is that it facilitates occasional control. This is particularly relevant in cases like the current situation (i.e., possible leak or until leak is arrested).

SOLUTION 5.18

Air Stripping and the Minimum Air to Water Ratio of the *NCEES Handbook* Section 6.9.4.4 is referred to as follows:

$$\left(\frac{Q_a}{Q}\right)_{min} = \frac{C_o - C_e}{HC_o}$$

$$= \frac{C_o - 0.5 \times C_o}{0.138 \times C_o}$$

$$= 1.45 \times 3.62$$

$$\left(\frac{Q_a}{Q}\right)_{optimal} = 7 \times 3.62 = 25.4$$

$$Q_a = 25.4 \times 2\ MGD \times 1.547\ \frac{cfs}{MGD}$$

$$= 78.6\ cfs\ (*)$$

Correct Answer is (D)

(*) Normally this value is optimized with the use of pilot tests to determine the final optimal value for air flow per contaminant.

Moreover, reference should also be made to the *Recommended Standards for Water Works Facilities, 2018,* Section 4.7.5.1 (d) for a packed tower aeration process design where it states that the volumetric air to water ratio at peak flow should be a minimum of 25 and a maximum of 80.

SOLUTION 5.19

Total hardness is caused by polyvalent cations, most predominantly calcium Ca^{2+} and magnesium Mg^{2+}.

Total hardness is divided into:

- Carbonate Hardness (CH): cations in this type are associated with carbonate CO_3^{2-} or bicarbonate HCO_3^-. Both of those anions represent alkalinity as well. This type of hardness can be removed by heat.

- Noncarbonate Hardness (NH): cations in this type are associated with ions such as Cl^-, SO_4^{2-} and NO_3^-. This type is the permanent type.

Applying the above definition to the given table can lead to the following calculations:

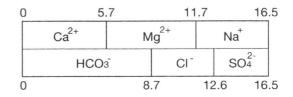

➢ *First row of the table – Cations:*

- Total Hardness (TH) is represented by concentrations of Ca^{2+} and Mg^{2+}:

$$TH = 11.7\ MEQ/L$$

➢ *Second row of the table – Anions:*

- Carbonate Hardness (CH) is represented by HCO_3^- when associated with the hardness cations of the first row:

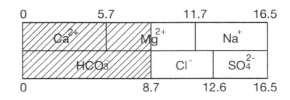

$$CH = 8.7\ MEQ/L$$

- Noncarbonate Hardness (NH) is represented by the other anions (Cl^-, SO_4^{2-} or NO_3^-) when associated with the hardness cations of the first row:

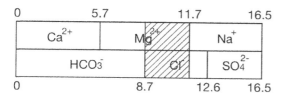

$$NH = 11.7 - 8.7 = 3\ MEQ/L$$

To convert this value into $mg/L\ CaCO_3$, multiply it by the equivalent weight of $CaCO_3$ which is '50'. This weight is also available in Section 6.9.5.2 Common Water Softening Compounds and Molecular Properties of the *NCEES Handbook*

$$NH = 3 \times 50 = 150\ mg/L\ CaCO_3$$

Correct Answer is (A)

SOLUTION 5.20

Reference is made to the *NCEES Handbook* Section 6.9.6 Settling and Sedimentation. Also, since the table given in the question implies the settlement of discrete particles, equations from Section 6.9.6.3 Type 1 – Discrete Particle Settling are used as follows:

$$v_o = \frac{Q}{WL}$$

$$= \frac{4.5\ MGD \times 1.547\ \frac{cfs}{MGD}}{25\ ft \times 75\ ft}$$

$$= 3.71 \times 10^{-3}\ ft/sec\ (= 13.37\ ft/hr)$$

Where v_o is the overflow rate, or putting it in other words, it is the speed by which the water surface moves up.

The particle settling velocity is v_t, and for sedimentation to occur → $v_t \geq v_o$. The removal ratio per particle in this case is calculated as follows which is presented in column II of the following table:

$$r = \frac{v_t}{v_o}$$

I	II	III	II x III
Average Settling velocity (v_o)	Removal ratio (fraction of particles removed)	Number of particles	Number of particles removed
ft/hr	(r)	per mL	per mL
3	0.224	1,300	291
6	0.449	1,800	808
9	0.673	2,000	1,346
12	0.898	1,100	988
Total	Total	6,200	3,433

Based on the above table, removal efficiency is calculated as follows:

$$efficiency = \frac{3,433}{6,200} = 55.4\%$$

Correct Answer is (C)

SOLUTION 5.21

There are two key observations in this given graph:

○ Excess Water Supply: When the flow demand line (representing consumption) falls below the provided flow line at a specific hour of the day, it indicates an excess in water supply. Without a properly sized distribution reservoir, this excess could lead to facility flooding.

○ Supply Deficit: Conversely, when the pumped flow exceeds the consumed demand, it indicates a deficit in supply.

To address these situations, we calculate the maximum deficit in supply (measured as $3\ million\ Liters$) and the excess in supply (measured as $1.2\ million\ Liters$). The required volume is obtained by adding these two values together.

$$V = 3 + 1.2 = 4.2\ million\ Liters$$

$$= 148,321.6\ ft^3$$

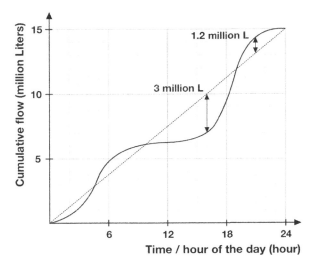

Normally, a factor of safety is added to this volume along with fire storage and any expected breakdowns. Fire and other requirements were excluded in this question.

Adding a factor of safety of 25% as requested in the questions:

$$V = 185,402 \, ft^3$$

Checking the volume of the provided options per the following table:

Option	Dimensions	Volume
A	95 ft dia 18 ft deep	127,587.9 ft^3
B	105 ft dia 18 ft deep	155,862.3 ft^3
C	110 ft dia 20 ft deep	190,066.4 ft^3
D	110 ft dia 30 ft deep	285,099.5 ft^3

Based on the above, Option C is considered the best option.

Correct Answer is (C)

SOLUTION 5.22

Compounding growth is modeled using the percent growth model equation found in the *NCEES Handbook* Section 6.1.6.3 as follows:

$$P_t = P_o(1 + k)^n$$

$$165,000 = 75,000(1 + k)^{20}$$

$$\rightarrow k = \left(\frac{165,000}{75,000}\right)^{1/20} - 1 = 4\%$$

Consumption of 7.5 MGD can serve the following population:

$$P = \frac{7.5 \, MGD}{5 \, MGD} \times 75,000 = 112,500$$

A population of 112,500 is reached in n years using the percent growth model as follows:

$$112,500 = 75,000(1 + 0.04)^n$$

$$1.5 = (1.04)^n$$

Using the mathematical equations and identifiers from the *NCEES Handbook's* first chapter:

$$n = \frac{\ln(1.5)}{\ln(1.04)}$$

$$n = 10.3$$

Correct Answer is (C)

📖 SOLUTION 5.23

Reference is made in this question to the *NCEES Handbook* Section 6.9.8 Membrane Filtration/Operations/Rate of Rejection.

The following sketch explains some of the terminologies used in this question:

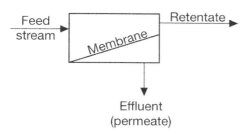

The feed stream is the influent with a concentration of $C_f = 2,000 \; mg/L$.

The effluent is the permeate stream with a concentration of $C_p = 75 \; mg/L$.

The stream that is left behind is the retentate – also referred to as the concentrate stream. This stream has a concentration of $C_c = 15 \; kg/m^3 \, (15,000 \; mg/L)$.

Based on the above, the concentration gradient can be calculated as follows:

$$\Delta C_i = \left[\frac{C_f + C_c}{2}\right] + C_p$$

$$= \left[\frac{2,000 + 15,000}{2}\right] + 75$$

$$= 8,575 \; mg/L \; (8.6 \; kg/m^3)$$

Correct Answer is (D)

📖 Concentration gradient is the difference in solute concentration between two regions separated by a membrane. The solute moves from the area of higher concentration to an area of lower concentration – also known as diffusion. The concentration gradient is the driving force that moves solutes across the membrane.

📖 **SOLUTION 5.24**

Head losses through a clean filter bed is discussed in the *NCEES Handbook* Section 6.9.7.

Given the sand in this filter is uniform, the monosized media equation is used and the *Rose* equation of Section 6.9.7.2 is applied here considering the information given in the question:

$$h_f = \frac{1.067(v_s)^2 \, L \, C_D}{\Psi g \eta^4 d}$$

$$\to L = \frac{h_f \, \Psi g \eta^4 d}{1.067(v_s)^2 \, C_D}$$

Where:

h_f is the given head loss = $3.5 \; ft$

Ψ is shape factor = 0.85

g is the gravitational acceleration = $32.174 \; ft/sec^2$

η is porosity and is given in the question as '0.45'

d is the diameter of the media particles and is given in the question as $0.0018 \; ft$

v_s is the loading rate calculated as follows:

$$= \frac{Q}{A_{plan}}$$

$$= \frac{0.5 \; MGD \times \frac{1.547 \; ft^3/sec}{MGD}}{100 \; ft^2}$$

$$= 7.735 \times 10^{-3} \; ft/sec$$

C_d is the drag coefficient and it can be determined with reference to Section 6.9.6.1 where *Reynolds* number (R_e) determines which C_d equation shall be used. See below:

$$R_e = \frac{v\rho d}{u}$$

$$= \frac{7.735 \times 10^{-3} \frac{ft}{sec} \times \frac{62.3 \frac{lbm}{ft^3}}{32.174 \frac{lbm.ft}{lbf.sec^2}} \times 0.0018\, ft}{2.05 \times 10^{-5} \frac{lbf.sec}{ft^2}}$$

$$= 1.315$$

Few important considerations for the above calculation:

- Viscosity u and water density ρ are taken from Section 6.2.1.6 for temperature of $70°\,F$
- Density ρ is divided by the gravitational acceleration to convert the pound mass lbm in the density unit into pound force lbf, which generates a unitless R_e
- $R_e > 1$ which means flow is transitional and C_d is calculated using the following equation:

$$C_d = \frac{24}{R_e} + \frac{3}{(R_e)^{1/2}} + 0.34$$

$$= \frac{24}{1.315} + \frac{3}{(1.315)^{1/2}} + 0.34$$

$$= 21.2$$

Substitute all the above into the first equation as follows:

$$L = \frac{h_f\, \Psi g \eta^4 d}{1.067 (v_s)^2\, C_D}$$

$$= \frac{3.5\, ft \times 0.85 \times 32.174 \frac{ft}{sec^2} \times (0.45)^4 \times 0.0018\, ft}{1.067 \times \left(7.735 \times 10^{-3} \frac{ft}{sec}\right)^2 \times 21.2}$$

$$= 5.2\, ft$$

Correct Answer is (D)

📖 *Rose* equation applied in this question has broader applicability compared to the *Carman-Kozeny* equation. The latter is primarily used in laminar flows, which is not the case for this specific question, as indicated by its *Reynolds* number (R_e).

Furthermore, the *Carman-Kozeny* equation does not consider the spherical shape of the media, making it more suitable for specific scenarios. In contrast, the *Rose* equation has wider applicability.

📖 SOLUTION 5.25

When *chlorine* is added to pure water, it reacts with it to form *hypochlorous acid* $HOCl$ and *hydrochloric acid* HCl.

$$Cl_2 + H_2O \rightarrow HOCl + HCl$$

When *hypochlorous acid* $HOCl$ dissociates in water, it forms the *hypochlorite ion* OCl^- which is another form of free *chlorine*.

$$HOCl \leftrightarrows H^+ + OCl^-$$

This reaction goes both ways and is pH dependent – i.e., When pH is low, the equilibrium shifts favoring the formation of hypochlorous acid $HOCl$ which is a more effective acid compared to hypochlorite ion OCl^-.

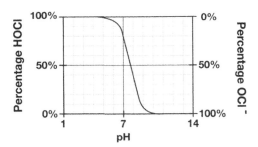

Since there are no impurities in pure water, it is expected that *chlorine* residual in water will equal to the added *chlorine* dose. For example, if we add a dose of $5\,mg/L$ of *chlorine*, it is then expected to have a $5\,mg/L$ *chlorine* residual in the form of either *hypochlorous acid* $HOCl$ or *hypochlorite ion*

OCl^- depending on pH as explained earlier. See below:

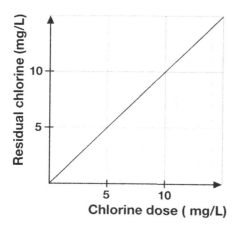

However, this is different when *chlorine* is added to impure water, such as groundwater or surface water. These waters contain impurities such as *iron, manganese,* or *ammonia*, all of which are called reducing agents. Those impurities are the result of the decaying natural organic matter.

With such impure waters, adding *chlorine* will not result in any *chlorine* residual, at least at the beginning – i.e., between points 1 and 2 of the curve below. *Chlorine* residual between these points is consumed by the reaction with the reducing agents mentioned earlier.

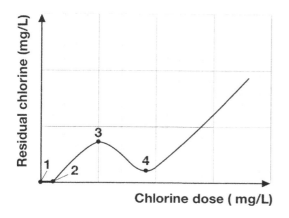

As we continue adding chlorine beyond point 2, chlorine continues reacting with the natural organic matter and the *ammonia* in the water to form *monochloramine* (predominantly) along with other compounds such as *trihalomethanes* and other chloro-organic compounds.

At the peak of point 3, and as we continue adding more *chlorine*, *chlorine* starts destroying these compounds until it bottoms out at the breakpoint – point 4.

Now as we continue adding more *chlorine* beyond point 4, a linear relationship forms between the increase in the *chlorine* dose and its residual.

Section 6.9.10 of the *NCEES Handbook* can be referred to for more clarity on the above.

The relevant statements of this question are underlined in the text for ease of reference. Based on this, the following statements in the question are the only true ones:

- Statement II
- Statement IV
- Statement V

Correct Answer is (A)

📖 SOLUTION 5.26

Reference is made in this question to the *NCEES Handbook,* Section 6.9.10.4 Chlorine Contact Chambers.

First start with determining CT_{actual}:

Compute the hydraulic residence time θ which is the time that *chlorine* stays in the tank and is calculated as follows:

$$\theta = \frac{5,000\ gallons}{150\ gallon/minute} = 33.33\ min$$

$$t_{10} = \theta \times BF = 33.33 \times 0.7 = 23.33\ min$$

$$CT_{actual} = C \times t_{10}$$
$$= 1.2\ mg/L \times 23.33\ min$$
$$= 28.0\ mg.min/L\ (*)$$

$CT_{required}$ for 3-Log inactivation:

Use table in page 486 of the *NCEES Handbook version 2.0* (different handbook versions may have this table on a different page).

Select the block with temperature $20°\ C$, at $pH = 7.0$ and *chlorine* concentration of $1.2\ mg/L$. See below snippet of the table for ease of reference:

Chlorine Concentration (mg/L)	Temperature = 20 C pH						
	<=6.0	6.5	7.0	7.5	8.0	8.5	9.0
<=0.4	36	44	52	62	74	89	105
0.6	38	45	54	64	77	92	109
0.8	39	46	55	66	79	95	113
1.0	39	47	56	67	81	98	117
1.2	40	48	57	69	83	100	120
1.4	41	49	58	70	85	103	123
1.6	42	50	59	72	87	105	126
1.8	43	51	61	74	89	106	129
2.0	44	52	62	75	91	110	132
2.2	44	53	63	77	93	113	135
2.4	45	54	65	78	95	115	138
2.6	46	55	66	80	97	117	141
2.8	47	56	67	81	99	119	143
3.0	47	57	68	83	101	122	146

$$CT_{required} = 57\ mg.min/L$$

$$\frac{CT_{actual}}{CT_{required}} = \frac{28}{57} = 0.5$$

It is imperative that this value is well above '1.0' to ensure successful inactivation. If the required level of inactivation is not achieved, either increasing the chlorine dosage shall be considered, or improving the system shall take place by enhancing baffling mechanisms and incorporating specific processes to leverage the removal credits outlined in Section 6.9.12.

Correct Answer is (A)

(*) Version 2.0 of the *NCEES Handbook* presents the unit for CT as $mg.mm/L$ when it should be $mg.min/L$

📖 Log inactivation or removal system explained:

The log removal system is a way of measuring the effectiveness of a water treatment process in removing contaminants. It is based on a logarithmic scale, where each log unit represents a tenfold reduction in the concentration of the contaminant. For example, reducing a concentration of $1,000\ mg/L$ to $10\ mg/L$ is considered a 2-Log removal because it involves a reduction of 100 times or 10^2.

On the other hand, the percentage removal system is based on the percentage of the contaminant that is removed from the water. With this system, reducing a concentration of $1,000\ mg/L$ to $10\ mg/L$ is considered a $\left(\frac{1,000-10}{1,000} = 99\%\right)$ removal.

See below table for some comparisons:

Log reduction	Percent reduction
1-Log	90%
2-Log	99%
3-Log	99.9%
4-Log	99.99%

While both systems are used to measure the effectiveness of water treatment processes, the log removal system provides a better representation of concentration reduction. The percentage removal system on the other hand can be misleading.

SOLUTION 5.27

The *Recommended Standards for Water Works Facilities, 2018,* Section 4.4.2 Residual Chlorine, states that the minimum free *chlorine* residual in water distribution systems should be 0.2 mg/L.

Correct Answer is (C)

SOLUTION 5.28

The *Recommended Standards for Water Works Facilities, 2018,* Section 9.3 (a) Precipitative Softening Sludge (Lagoons), states that short term lagoons area in *acres* should be calculated as follows (a usable depth of 5 ft should be considered as well):

$$= 0.7 \times \frac{flow\ (MGD)}{1\ MGD} \times \frac{Hardness\ removed\ (mg/L)}{100\ mg/L}$$

$$= 0.7 \times \frac{7\ MGD}{1\ MGD} \times \frac{150\ mg/L}{100\ mg/L}$$

$$= 7.35\ acres$$

The above section also states that it is preferable to provide this area over two or more lagoons.

Based on the above, the following options are the only workable options for this scenario:

- ☑ Two lagoons each with an area of 4.0 *acres* along with 5 ft usable depth.
- ☑ Four lagoons each with an area of 2.0 *acres* along with 5 ft usable depth.

The rest of the options either do not provide the required area or they do not provide the required usable depth.

SOLUTION 5.29

The activated sludge system is a common wastewater treatment process that consists of an aeration basin and a clarifier (settling basin). The flow enters the aeration basin, where it is aerated with oxygen and air to support the growth of microorganisms that degrade organic matter. The microorganisms settle to the bottom of the clarifier, where they are recycled back to the aeration basin to continue their work on the contaminants. See below sketch:

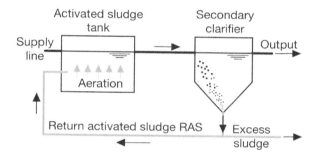

Aeration is necessary because the number of microorganisms requires extra oxygen beyond what can be dissolved through the atmosphere. However, it is important to maintain the dissolved oxygen concentration within certain operating parameters. Typically, the concentration of oxygen should be around 2 mg/L (*) to effectively degrade the organic matter in the waste. If the concentration of oxygen is too high, energy will be wasted, making the process inefficient and expensive. Additionally, high oxygen concentrations can result in the growth of *filamentous* bacteria, which do not settle like other organisms in the clarifier, rendering the reactivation process ineffective. Conversely, if the concentration of dissolved oxygen is too low, the process becomes anaerobic, leading to odor problems.

Based on the above, the following statements are the only accurate ones: II + III

Correct Answer is (B)

(*) See section 92.333 (b) Aeration and the Mechanical Aeration Systems from the *Recommended Standards for Wastewater Facilities, 2014* for more details on this.

SOLUTION 5.30

Aeration should be provided to equalization tanks to prevent septicity and solids build-up. It is stated in the *Recommended Standards for Wastewater Facilities, 2014,* Section 65.52 Aeration (flow equalization), that a level of $1\ mg/L$ of dissolved oxygen shall always be maintained.

Correct Answer is (B)

SOLUTION 5.31

Below is a high-level description for each of the presented processes and their use:

Activated Sludge Process (ASP):
Utilizes microorganisms to break down organic matter; requires aeration tanks and clarifiers. Common for achieving high levels of organic matter removal, and effective in treating organic-rich industrial wastewater. It requires a large area for implementation.

Trickling Filter:
Involves trickling wastewater over a bed of rocks or plastic media for microbial growth. Used for smaller Treatment Facilities, also suitable for decentralized systems and areas with space constraints. It is effective for organic matter reduction.

Membrane Bioreactor (MBR):
Integrates biological treatment with membrane filtration. It is a compact solution commonly used in urban areas. The effluent is of a higher quality that is suitable for reuse.

Rotating Biological Contactor (RBC):
Consists of rotating disks that support microbial growth and facilitates biological treatment. Effective for small to medium plants and can be implemented vertically if needed.

Based on the above description, matching can be done as follows:

Location size and requirements	Secondary Treatment Process
Urban areas with compact solutions and the need for a high-quality effluent for reuse.	Membrane Bioreactor (MBR)
Needed for efficient organic matter content removal, space is not an issue.	Activated Sludge Process (ASP)
Needed mostly for organic matter reduction, decentralized system, and area with space constraints.	Trickling Filter
Needed for moderate organic load removal and a limited space for the system. It may need to be installed vertically.	Rotating Biological Contactor (RBC)

SOLUTION 5.32

Determine the suspended solids weight (or loading – i.e., SL) per day for the RAS line using the *NCEES Handbook* Section 6.8.5.3 Activated Sludge Treatment. The Solid Loading equation of page 447 of the handbook's version 2.0 is used as follows:

$$SL = 8.34\ QX$$

$$= 8.34 \times 5\ MGD \times 7{,}500\ mg/L$$

$$= 312{,}750\ lb/day$$

The *chlorine* weight is calculated as a percentage of this value as follows:

$$Weight_{Cl_2} = 312{,}750\ \frac{lb}{day} \times 0.5\%$$

$$= 1{,}563.75\ \frac{lb}{day}$$

Correct Answer is (B)

SOLUTION 5.33

Denitrification is the process of converting *nitrate* and *nitrite* into *nitrogen* gas. This process is achieved with the use of *heterotrophic* bacteria which is the bacteria that uses organic carbon as their food source, hence the addition of *methanol* (CH_3OH). See solution of question 5.40 for more details on this process.

These bacteria can break *nitrate* (NO_3^-) and/or *nitrite* (NO_2^-) molecules to use oxygen and then release *nitrogen* as gas.

The *NCEES Handbook*, page 452 of its 2nd version, presents an equation that computes the required concentration of *methanol*:

$$C_m = 2.47\ (NO_3 - N) + 1.53(NO_2 - N) + 0.87 DO$$

$$= 2.47 \times 25 + 1.53 \times 0 + 0.87 \times 8$$

$$= 68.71\ mg/L$$

Convert this value into the required solids loading using the *NCEES Handbook* Section 6.8.5.3 Activated Sludge Treatment, Solid Loading equation in page 447:

$$SL = 8.34\ QX$$

$$= 8.34 \times 0.5\ MGD \times 68.71\ mg/L$$

$$= 286.52\ lb/day$$

Correct Answer is (C)

SOLUTION 5.34

Reference is made to Section 42.76 of the *Recommended Standards for Wastewater Facilities, 2014* where the requirement for pumping stations' dry well continuous ventilation is stated as six complete air changes per hour (ACH).

In order to convert this into CFM, volume of the dry well is used as follows:

$$= 6\ \frac{AC}{hr} \times \frac{(7 \times 6 \times 8)\ ft^3 per\ AC}{60\ \frac{min}{hr}}$$

$$= 33.6\ CFM$$

Correct Answer is (A)

SOLUTION 5.35

The height of the cascade aerator can be calculated using the *NCEES Handbook* Section 6.8.11.1 Cascade Aeration using the following equation:

$$H = \frac{R - 1}{0.11 ab(1 + 0.046\ T)}$$

$$R = \frac{C_s - C_o}{C_s - C}$$

Where T is temperature in Celsius, and C represents the concentration of Dissolved Oxygen (DO) as follows:

Temp	Temp	Oxygen Solubility (mg/L)	Initial DO (mg/L)	Desired DO (mg/L)
F	C	C_s	C_o	C
59°	15°	10.084	2.0	6.0
86°	30°	7.559	2.0	6.0

Height at 15° C:

$$R = \frac{C_s - C_o}{C_s - C}$$

$$= \frac{10.084 - 2.0}{10.084 - 6.0}$$

$$= 1.98$$

$$H = \frac{R-1}{0.11ab(1+0.046\,T)}$$

$$= \frac{1.98 - 1}{0.11 \times 0.8 \times 1.1(1+0.046 \times 15)}$$

$$= 6.0\,ft$$

Height at 30° C:

$$R = \frac{C_s - C_o}{C_s - C}$$

$$= \frac{7.559 - 2.0}{7.559 - 6.0}$$

$$= 3.57$$

$$H = \frac{R-1}{0.11ab(1+0.046\,T)}$$

$$= \frac{3.57 - 1}{0.11 \times 0.8 \times 1.1(1+0.046 \times 30)}$$

$$= 11.2\,ft$$

As the temperature rises, the deficit ratio (R) also increases. Consequently, when the temperature is higher, the step cascade system height should be increased.

Height is therefore calculated as follows:

$$H = max(11.2\,ft, 6.0\,ft) = 11.2\,ft$$

Hence, the number of steps required using a $1\,ft$ riser per step is 12 steps.

Correct Answer is (D)

(✻) **SOLUTION 5.36**

Start with determining the sludge mass that is transferred from the primary clarifier to the digester as follows ($MG = Million\,Gallon$):

$$= 8.34\,QX$$

$$= 8.34\,Q \times (TSS \times 45\%\,removal)$$

$$= \left(8.34\,\frac{lb}{MG.mg/L}\right) \times 5\,\frac{MG}{day} \times 250\,mg/L \times 0.45$$

$$= 4{,}691.25\,lb/day$$

It is important to understand that when a primary clarifier removes 45% of TSS, the resulting sludge (i.e., the 45% of Total Suspended Solids removed) is transferred to the digester and the rest goes to secondary treatment. Same applies to BOD reduction.

Sludge volume is calculated using its specific gravity '1.03' multiplied by the water specific weight $62.4\,lb/ft^3$ after removing the moisture content of 94% - see page 449 of the handbook's 2nd version:

$$V_{sludge} = \frac{4{,}691.25\,lb/day}{1.03 \times 62.4\,lb/ft^3 \times (1 - 0.94)}$$

$$= 1{,}216.5\,ft^3/day$$

The digester volume is computed using the Solid Residence Time SRT – reference to the *NCEES Handbook* Section 6.8.9.2 can be made:

$$V_{digester} = 1{,}216.5\,ft^3/day \times 20\,days$$

$$= 24{,}330.2\,ft^3$$

Volume of *methane* (CH_4) can be determined as 0.35 *Liters* for each *gram* of BOD, corrected to the amount of BOD in influent/effluent difference along with the net mass of cell tissues produced P_x – see *NCEES Handbook* Section 6.8.9.2 for details.

$$V_{CH_4} = 0.35 \frac{L_{CH_4}}{g_{BOD}} \left| (S_o - S)Q - 1.42 \frac{g_{BOD}}{g_{biomass}} P_x \right|$$

$$= 0.35 \frac{L_{CH_4}}{g_{BOD}} \left| (QS_o - QS) - 1.42 \frac{g_{BOD}}{g_{biomass}} P_x \right|$$

Based on the above, the following parameters can be calculated:

The influent BOD to the digestor:

$QS_o = 8.34 \, QX$

$= 8.34 \, Q \times (BOD \times 35\%)$

$= \left(8.34 \frac{lb}{MG.mg/L} \right) \times 5 \frac{MG}{day} \times 350 \, mg/L \times 0.35$

$= 5{,}108.25 \, lb/day$

The effluent BOD from the digestor:

$QS = (1 - 0.7)QS_o$

$= (1 - 0.7) \times 5{,}108.25 \, lb/day$

$= 1{,}532.5 \, lb/day$

The second part of the *methane* volume equation is calculated as follows:

$= \left| (QS_o - QS) - 1.42 \frac{g_{BOD}}{g_{biomass}} P_x \right|$

$= |(5{,}108.25 - 1{,}532.5) - 1.42 \times 89.3|$

$= 3{,}449 \, lb/day$

Volume of *methane* (*) per day is therefore calculated as follows – pay close attention to units' conversions:

$V_{CH_4} = 0.35 \frac{L_{CH_4}}{g_{BOD}} \times \frac{0.0353 \frac{ft^3}{L}}{2.205 \times 10^{-3} \frac{lb}{g}} \times 3{,}449 \frac{lb}{day}$

$= \mathbf{19{,}325.2 \, ft^3/day}$

(*) It is also important to recognize that in both standard-rate and high-rate digesters, *methane* constitutes 65% of total gas volume. This is also mentioned in Section 6.8.9.2 of the handbook.

$$V_{total \, gases} = \frac{19{,}369.3}{0.65} = 29{,}798.9 \, ft^3/day$$

SOLUTION 5.37

Scour velocity (v_c) is the flow velocity in the tank beyond which it causes resuspension (scouring) of the settled particles, this is why the horizontal velocity throughout settling tanks shall be kept well below the scour velocity – refer to the *NCEES Handbook* Section 6.8.5.2 Clarifiers, Scour Velocity on page 445 of version 2.0 for more details:

$$v_c = \left[\frac{8 \beta (s-1) g d}{f} \right]^{0.5}$$

$$= \left[\frac{8 \times 0.05 (1.25 - 1) \times 32.2 \frac{ft}{sec^2} \times 3.28 \times 10^{-4} \, in}{0.025} \right]^{0.5}$$

$$= 0.206 \, ft/sec$$

The question states that a factor of safety of '2' shall be used, which means that selected tanks' cross-sections shall be designed to achieve half of this velocity.

Also, it is important to remember that the minimum side water depth per the *Recommended Standards for Wastewater Facilities, 2014* Chapter 70 Settling, Section 72.1 Dimensions, is $10 \, ft$. This eliminates options (1) and (5).

Using the velocity equation $Q = vA$, widths (*) for options (2), (3) and (4) can be determined against their proposed heights and then checked if they satisfy half the scour velocity as follows:

$$W = \frac{Q}{2 \text{ clarifiers} \times \left(\frac{v_c}{2}\right) \times H}$$

Option (2) with $H = 10\ ft$:

$$W = \frac{\left(30\ MGD \times \frac{1.547\ ft^3/sec}{MGD}\right) ft^3/sec}{2 \times \left(\frac{0.206\ ft/sec}{2}\right) \times 10\ ft}$$

$$= 22.5\ ft < 24\ ft \rightarrow OK$$

Option (3) with $H = 10\ ft$:

$$W = \frac{\left(30\ MGD \times \frac{1.547\ ft^3/sec}{MGD}\right) ft^3/sec}{2 \times \left(\frac{0.206\ ft/sec}{2}\right) \times 10\ ft}$$

$$= 22.5\ ft > 20\ ft \rightarrow NOT\ OK$$

Option (4) with $H = 12\ ft$:

$$W = \frac{\left(30\ MGD \times \frac{1.547\ ft^3/sec}{MGD}\right) ft^3/sec}{2 \times \left(\frac{0.206\ ft/sec}{2}\right) \times 12\ ft}$$

$$= 18.8\ ft < 20\ ft \rightarrow OK$$

Correct Answers (2) & (4)

(*) Alternatively, velocity for each section can be computed using $v = Q/A$ taking into account that there are two tanks not one, and then, compared to half of the scour velocity.

SOLUTION 5.38

The *NCEES Handbook* Section 6.8.5.6 Biotowers/Trickling Filter, and the Trickling Filter Performance equation found in page 455 of the handbook version 2.0 can be referred to as follows:

$$E_1 = \frac{100}{1 + 0.0561 \sqrt{\frac{W}{VF}}}$$

BOD loading to filter:

$$W = 8.34 QX$$

$$= \left(8.34\ \frac{lb}{MG \cdot mg/L}\right) \times 0.2\ \frac{MG}{day} \times 250\ mg/L$$

$$= 417\ lb/day$$

Trickling filter volume in thousand ft^3:

$$V = \pi \times (30\ ft)^2 \times 12\ ft$$

$$= 33{,}929.2\ ft^3$$

$$= 33.93\ \text{thousand}\ ft^3$$

Recirculation factor:

$$F = \frac{1 + \frac{R}{I}}{\left(1 + 0.1 \times \frac{R}{I}\right)^2}$$

$$= \frac{1 + \frac{1\ cfs}{\left(0.2\ MGD \times \frac{1.547\ ft^3/sec}{MGD}\right) cfs}}{\left(1 + 0.1 \times \frac{1\ cfs}{\left(0.2\ MGD \times \frac{1.547\ ft^3/sec}{MGD}\right) cfs}\right)^2}$$

$$= 2.42$$

Performance:

$$E_1 = \frac{100}{1 + 0.0561 \sqrt{\frac{W}{VF}}}$$

$$= \frac{100}{1 + 0.0561 \sqrt{\frac{417}{33.93 \times 2.42}}}$$

$$= 88.8\ \%$$

Influent BOD removal:

$$S_e = (1 - E_1) S_o$$

$$= (1 - 0.888) \times 250$$

$$= 28\ mg/L$$

Correct Answer is (A)

SOLUTION 5.39

The question refers to the air flotation process achieved via a Dissolved Air Flotation (DAF) system.

Flotation is the process of separating solids and liquid with the introduction of air usually. Air bubbles are attached to solids and through buoyancy they rise to the surface. This is achieved by dissolving air into wastewater under pressure. Normally the entire flow is held in a tank under pressure for several minutes until air is dissolved, then released into the flotation tank.

The equation to calculate the optimum air to solids ratio along with the required pressure is presented in the *NCEES Handbook* Section 6.8.5.4 as follows:

$$\frac{A}{S} = \frac{1.3 s_a (fP - 1)}{S_a}$$

Where S_a (with a capital 'S') is the influent suspended solids' concentration, given in the question as 0.5%. Refer to the below conversion calculator:

$$1\% \ solids = 10,000 \ mg/L = 10,000 \ PPM$$

$$\rightarrow S_a = 0.5\% = 5,000 \ mg/L$$

s_a (with a small 's') is air solubility, which can be picked up from the same section of the handbook as 15.7 mL/L at $30°\ C$.

f is the fraction of air dissolved at pressure P and is usually taken as '0.5'. And p is the gauge pressure which will be calculated later.

$$\frac{A}{S} = \frac{1.3 s_a (fP - 1)}{S_a}$$

$$0.008 = \frac{1.3 \times 15.7 \ mL/L \times (0.5 \times P - 1)}{5,000 \ mg/L}$$

$$P = 2 \times \left(\frac{0.008 \times 5,000}{1.3 \times 15.7} + 1\right) = 5.92 \ atm$$

In order to convert this into gauge pressure in kPa, the following equation is used, which is also provided in the same section of the handbook:

$$P = 5.92 \ atm = \frac{p + 101.35}{101.35}$$

$$\rightarrow p = 498.64 \ kPa$$

Correct Answer is (C)

📖 SOLUTION 5.40

In an activated sludge process, treatment is accomplished by removing BOD and *ammonia* (NH_3) from wastewater. The remaining flow which goes into the clarifier contains *nitrate* (NO_3^-) that is generated due to the removal of *ammonia*. See below:

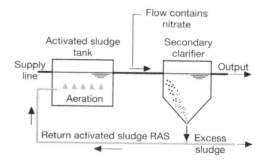

The removal of *nitrate* from this flow is achieved with the addition of a zone prior to the clarifier called the anoxic zone. The anoxic process, similar to an anaerobic process, does not require dissolved oxygen to work. However, dissimilar to the anaerobic process, the anoxic process requires the *heterotrophic* bacteria to consume the oxygen molecule attached to the *nitrate* compound (NO_3^-). *Nitrogen* gas is released because of this, and denitrification is accomplished.

However, in order to achieve the above, carbon should be added to the anoxic zone in the form of *methanol* (CH_3OH) or *acetate* ($C_2H_3O_3^-$) which helps the bacteria consume the said oxygen molecule.

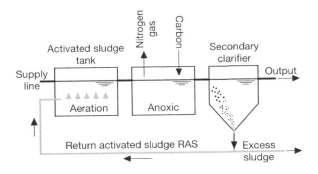

To get rid of the ongoing chemical cost represented by the addition of carbon (*methanol* or *acetate*), the process can be adjusted by starting with denitrification at the very beginning of the biological treatment – i.e., start with an anoxic zone to perform denitrification. At this stage of the process, the influent contains enough carbon to satisfy the anoxic reaction.

The flow then moves from the anoxic zone into the usual aerobic process for the removal of BOD and *ammonia*. *Nitrate* will form again due to the removal of *ammonia*. Hence, the flow after the aerobic zone is recirculated to the initial anoxic zone for further denitrification. See below:

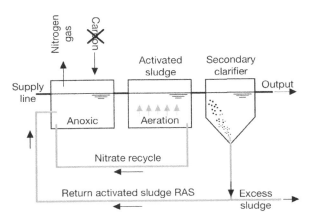

There has been one more improvement to this overall procedure which is referred to as the *Bardenpho* process. In this process, two more zones are added post the aerobic zone to get rid of residuals – i.e., after the aerobic zone, an anoxic zone is added, then another aerobic one. See below diagram:

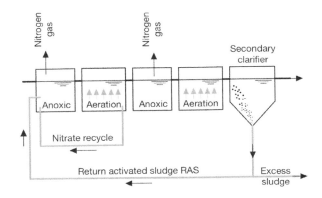

Based on the above, the best option which represents a proper biological denitrification is option (C).

It is also possible to make one further adjustment to this overall process to achieve biological *phosphorus* removal as well. See section below this answer for how.

Correct Answer is (C)

📖 Biological phosphorus removal:

Phosphorus can be removed biologically by adding an anaerobic zone to the wastewater treatment process. This process is accomplished by microorganisms called Phosphorus Accumulating Organisms (PAOs), which use *phosphorus* as their source of energy. In the absence of oxygen, PAOs consume *phosphorus* in the influent.

A simple modification to the above-described biological denitrification process can be made to achieve biological *phosphorus* removal by adding an anaerobic zone at the very start of the process.

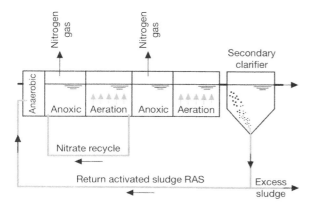

The above modification can reduce *phosphorus* to about $1\ mg/L$. If a plant has a requirement to reduce *phosphorus* to a level lower than that, chemicals have to be involved in the form of alum, *ferric chloride, ferrous chloride, ferric sulfate,* or *lime.* (See Section 6.8.7.2 of the *NCEES Handbook version 2.0* for the reaction equations).

It is important to know that excessive amounts of nutrients, primarily *nitrogen* and *phosphorus*, causes *eutrophication*, which is the growth of algae. Algae feeds on these nutrients, along with *carbon dioxide*, and spread across the water body turning it into green.

The process described above is quite complex and requires a more detailed explanation. However, this summary was provided to give you a high-level understanding of the available systems and why they have been developed as such.

SOLUTION 5.41
Reference can be made to the *NCEES Handbook version 2.0* Section 6.8.5.3 and page 447.

The formula of F/M is as follows – also please note that *MLVSS* should be used when available not *MLSS* as stated in the handbook:

$$F/M = \frac{S_o Q_o}{VX} = \frac{BOD\ (mass\ per\ day)}{MLVSS\ (avaiable\ in\ basin)}$$

MLVSS is the Mixed Liquor Volatile Suspended Solids concentration, and the term volatile references solids that are burnt and evaporated in a furnace. Given that the sample was baked in an oven, this value can be determined as follows:

$$MLVSS = \frac{Wt.of\ volatile\ solid}{Volume}$$

$$= \frac{(633.2 - 215.1)\ mg}{300 \times 10^{-3}\ L}$$

$$= 1{,}393.7\ mg/L$$

Although BOD and *MLVSS* values are available and can be incorporated in the F/M equation, volume of flow and volume of basin shall be incorporated as follows:

$$F/M = \frac{S_o Q_o}{VX}$$

$$= \frac{350\ \frac{mg}{L} \times 3.0 \times 10^6\ gallon\ per\ day}{200 \times 10^3\ ft^3 \times 7.463\ \frac{gal}{ft^3} \times 1{,}393.7\ mg/L}$$

$$= 0.5\ day^{-1}$$

Also, it is to be noted that this value can only be used the day the sample was collected assuming the sample is representative.

Correct Answer is (A)

SOLUTION 5.42

The flow in the Return Activated Solids' RAS (Q_R) line is calculated using the *NCEES Handbook version 2.0* – equation found in page 452 as follows (*):

$$Q_R = \frac{SV \times Q}{1{,}000 - SV}$$

SV is the volume occupied by $1{,}000\ mL$ of $MLSS$ after 30 *minutes* settling and it can be calculated using SVI (see page 449 of the handbook):

$$SVI = \frac{SV \times 1{,}000}{MLSS}$$

$$SV = \frac{SVI \times MLSS}{1{,}000}$$

SVI is calculated using the following equation (page 448 of the handbook):

$$TSS = \frac{\left(1{,}000\frac{mg}{g}\right)\left(1{,}000\frac{mL}{L}\right)}{SVI}$$

$$SVI = \frac{\left(1{,}000\frac{mg}{g}\right)\left(1{,}000\frac{mL}{L}\right)}{TSS}$$

$$= \frac{\left(1{,}000\frac{mg}{g}\right)\left(1{,}000\frac{mL}{L}\right)}{6{,}500\ mg/L}$$

$$= 153.85\ mL/g$$

$$SV = \frac{SVI \times MLSS}{1{,}000}$$

$$= \frac{153.85 \times 3{,}500}{1{,}000}$$

$$= 538.46\ mL/L$$

$$Q_R = \frac{SV \times Q}{1{,}000 - SV}$$

$$= \frac{538.46 \times 5}{1{,}000 - 538.46}$$

$$= 5.83\ MGD$$

(*) Alternatively, a shorter method can be used using the same equation of page 452 as follows:

$$Q_R = \frac{QX}{X_R - X}$$

Where X represents $MLSS$ concentration in the aerated tank and X_R represents the TSS concentration in the RAS line. See below:

$$Q_R = \frac{QX}{X_R - X}$$

$$= \frac{5 \times 3{,}500}{6{,}500 - 3{,}500}$$

$$= 5.83\ MGD$$

SOLUTION 5.43

Although chlorination is used to disinfect water prior to its discharge, disinfected water with significant levels of *chlorine* residuals is toxic to aquatic life. Moreover, *chlorine* reaction with some organic material can produce carcinogenic compounds – examples are *trihalomethanes* and *organochlorines*. Hence why dechlorination (or *chlorine* limiting) is regulated by government agencies.

Using the *Recommended Standards for Wastewater Facilities, 2014* Section 103.2 stipulates that *sodium thiosulfate* as a solution can be used to neutralize Cl_2 at a theoretical ratio of $0.56 : 1$ of $Na_2S_2O_3$ to Cl_2.

The concentration of Cl_2 that needs to be neutralized per the set regulation is:

$$X_{Cl_2} = 1.75\frac{mg}{L} - 0.5\frac{mg}{L} = 1.25\frac{mg}{L}$$

Loading can is calculated using SL equation from the handbook version 2.0 page 447:

$$SL = 8.34\,Q \times (0.56\,X_{Cl_2})$$

$$= \left(\frac{8.34\ lb}{MG.\frac{mg}{L}}\right) \times 10\,\frac{MG}{day} \times \left(0.56 \times 1.25\,\frac{mg}{L}\right)$$

$$= 58.38\,\frac{lb}{day}$$

Per Section 103.2 of the *Wastewater Facilities standards*, the calculated dose is theoretical, and it is recommended to add 10% to it as follows:

$$SL = 58.38 \times 1.1 = 64.2\ lb/day$$

SOLUTION 5.44

Odor in water bodies can arise from various sources, with the most prevalent being the overgrowth of algae. When algae die and decompose, they deplete the oxygen supply, hindering the breakdown of organic waste, consequently leading to unpleasant odors.

Water stagnation is another contributing factor to foul odors. When water remains still, algae and bacteria grow, and these bacteria break down waste materials, generating *carbon dioxide* and *hydrogen sulfide*, resulting in an odor similar to rotten eggs.

Additionally, the accumulation of scum at the bottom of ponds, which is mostly contributed by runoffs, can also cause water to stink.

Effective strategies to address these issues are:

o Aeration: reintroducing oxygen into the pond to prevent stagnation. This can be accomplished through aeration systems like fountains and submerged aeration systems.

Specialized Bacterial Blends: The introduction of specific bacterial blends capable of mitigating algae growth through denitrification is beneficial. Stormwater ponds typically contain elevated levels of *nitrogen* and *phosphorus*, which serve as nutrients for algae. Specially designed bacterial treatments can digest these nutrients before algae can, thus impeding their growth. Moreover, these bacteria can reach the pond's bottom to break down the sludge layer formed by runoff, reducing odor.

The above makes statements I and IV true.

Apart from the above, it is important to note that chemical treatments can also address sludge and algae issues, as well as reduce *nitrogen* and *phosphorus* levels. However, the use of chemicals should be considered a last resort due to their potential adverse effects on the environment, aquatic life, and other ecologically sensitive elements.

Moreover, while physically removing algae and debris from ponds is another viable strategy, it tends to be a short-term solution, especially for large ponds, and may entail substantial costs. Furthermore, this approach does not address the underlying causes of the problem.

The above marks statements II and III as incorrect.

Correct Answer is (C)

SOLUTION 5.45

The equation and the Streeter Phelps model of the *NCEES Handbook*, Section 6.7.4 Oxygen Dynamics, is used in this question.

The saturation value of dissolved oxygen at temperature $20°\ C$ is taken from the relevant table of this section as $9.17 \, mg/L$, and the *Churchill et al.* method for calculating the e based reaeration coefficient k_r is presented in Section 6.7.4.6 as follows:

$$k_r = \frac{11.57 v^{0.969}}{H^{1.673}}$$

$$= \frac{11.57 \times (2 \, ft/sec)^{0.969}}{(7.5 \, ft)^{1.673}}$$

$$= 0.78 \, day^{-1}$$

Based on the information given in the question, the initial oxygen deficit at the mixing zone can be determined as follows:

$$D_a = 9.17 \, mg/L - 3.5 \, mg/L$$

$$= 5.67 \, mg/L$$

The time at which minimum dissolved oxygen occurs can be determined as follows:

$$t_c = \frac{1}{k_r - k_d} \ln \left[\frac{k_r}{k_d} \left(1 - D_a \frac{(k_r - k_d)}{k_d L_a} \right) \right]$$

$$= \frac{1}{0.78 - 0.6} \ln \left[\frac{0.78}{0.6} \left(1 - 5.67 \frac{(0.78 - 0.6)}{0.6 \times 35} \right) \right]$$

$$= 1.18 \, day$$

Using the above duration along with the average velocity of the river, the distance to the critical deficit can be calculated as follows:

$$Distance = v \times t_c$$

$$= 2 \, \frac{ft}{sec} \times \frac{\frac{1 \, mile}{5,280 \, ft}}{\frac{1 \, day}{86,400 \, sec}} \times 1.18 \, day$$

$$= 38.6 \, mile$$

Correct Answer is (D)

SOLUTION 5.46

This question can be solved using mass balance, and can be solved with the direct application of the initial deficit equation of the *NCEES Handbook* Section 6.7.4.3 as well as follows:

$$D_a = DO_{stream} - \frac{Q_w DO_w + Q_{stream} DO_{stream}}{Q_w + Q_{stream}}$$

$$= 6.5 - \frac{45 \times 2.2 + 175 \times 6.5}{45 + 175}$$

$$= 0.9 \, mg/L$$

Correct Answer is (A)

📖 SOLUTION 5.47

Reference is made to the *NCEES Handbook*, Section 6.7.5 *Monod* Kinetics – Substrate Limited Growth, and the graph presented in Section 6.7.5.5.

It can be observed from the graph presented in Section 6.7.5.5, and *Monod's* equation, that the relationship between growth rate and substrate concentration is non-linear, based upon which, a trendline can be added to the given data set to represent its general pattern as follows:

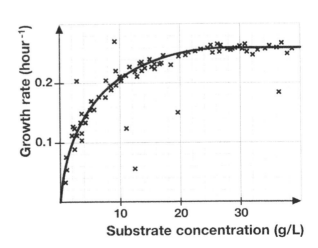

The above non-linear curve represents the following equation, where the maximum growth rate can be determined as $u_{max} = 0.26 \; hour^{-1}$ as noted from the following graph:

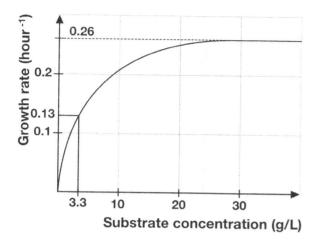

$$u = u_{max} \frac{S}{K_s + S} - k_d$$

In this equation, u is the growth rate, S is substrate concentrate in the solution, and k_d is the microbial death rate.

The saturation constant K_s requested in this question is the substrate concentration that occurs at half maximum growth (i.e., at $\frac{u_{max}}{2} = 0.13$) as shown in the above graph:

$$\rightarrow K_s = 3.3 \; g/L$$

Based on this, *Monod's* equation can be written as follows:

$$u = \frac{0.26 \, S}{3.3 + S} - k_d$$

Correct Answer is (C)

📖 *Monod's* model is a mathematical model that helps us understand how microorganisms grow. It tells us that the growth rate of microorganisms is proportional to the amount of food available (i.e., substrate), but it slows down as the food source becomes more abundant. This equation is used to predict how microorganisms will grow in wastewater treatment zones. It helps optimize conditions for microbial activity and understand how changes in food availability affect microbial growth.

In the context of wastewater treatment, microorganisms like *Escherichia Coli* or *Bacillus Subtilis* use organic matter like carbonaceous pollutants as a food source for their growth and metabolism. The *Monod* equation helps us understand how the concentration of organic matter affects the growth rate of these bacteria, which in turn influences the efficiency of the treatment process.

📖 SOLUTION 5.48

Reference is made to the *NCEES Handbook* Section 6.7.8.

Start with determining the reference dose RfD for this chemical.

RfD is the dose below which no adverse health effects should result from a lifetime exposure. A study is normally conducted where several doses are administered to animals. The RfD in this case is calculated by dividing the selected (normally the lowest) dose by an uncertainty factor to consider inadequacies of the study – i.e., animal-human extrapolation.

$$RfD = \frac{NOAEL}{uncertainty \; factor}$$

$$= \frac{0.3 \; mg/kg}{100}$$

$$= 0.003 \; mg/kg$$

The second step is to calculate the drinking water equivalent level (DWEL). In this step, the RfD is converted into a concentration in drinking water using the assumed consumption and body weight both of which were given in the question as follows:

$$DWEL = \frac{RfD \times body\ weight}{daily\ water\ consumption}$$

$$= \frac{0.003\frac{mg}{kg} \times 67\ kg}{2.3\frac{Liter}{day}}$$

$$= 0.087\ mg/L$$

The DWEL assumes that the daily (lifetime) exposure from a certain substance that comes through drinking water shall not have adverse health effects to humans. It is therefore used to determine Maximum Contaminant Level Goals MCLGs at later stage.

Correct Answer is (C)

📖 SOLUTION 5.49

Reference is made to the *NCEES Handbook* Section 6.7.8.5.

Start with determining the chronic daily intake CDI:

$$CDI = C\frac{(IR)(EF)(ED)}{(BW)(AT)}$$

$$= 0.02\frac{mg}{L} \times \frac{2\frac{L}{day} \times 365\frac{day}{yr} \times 40\ yr}{75\ kg \times 365\frac{day}{yr} \times 75\ yr}$$

$$= 0.000284\frac{mg}{kg.day}$$

CDI is used to calculate exposure limits for drinking water with the assumption that the acceptable level of risk is 10^{-5} to 10^{-6}. This means that CDI assumes that the acceptable likelihood of being diagnosed with cancer from drinking water with this level of contamination is one person in a population of 100,000 to 1,000,000 during a lifetime.

The second step is to calculate the risk by multiplying CDI with the given Slope Factor SF for the contaminant.

$$risk = CDI \times SF$$

$$= 0.000284\ \frac{mg}{kg.day} \times 0.35\ \frac{kg.day}{mg}$$

$$= 9.94 \times 10^{-5}$$

The Environmental Protection Agency EPA uses a linear model when it comes to calculating risk factors and slope factors SF. This means that there should be multiple events that took place, and the risk is assumed to decrease when exposure (i.e., dose) has decreased.

Correct Answer is (A)

SOLUTION 5.50

The only two correct statements are III and IV.

The corrected statements that should replace I and II are as follows:

I. TMDL is the calculation of the maximum amount of a pollutant allowed to enter a waterbody so that the waterbody will continue to meet water quality standards for that pollutant.

II. The objective of a TMDL is to determine the loading capacity of the waterbody and to allocate that load among different pollutant sources.

Correct Answer is (C)

VI
PROJECT SITE WORK

Knowledge Areas Covered

SN	Knowledge Area
12	**Project Sitework** A. Excavation and embankment (e.g., grading, cut and fill) B. Construction site layout and control C. Temporary and permanent soil erosion and sediment control (e.g., construction erosion control and permits, sediment transport, channel/outlet protection) D. Impact of construction on adjacent facilities E. Safety (e.g., construction, roadside, work zone) F. Basic horizontal and vertical curve elements G. Retaining walls H. Construction methods

PART VI
Project Site Work

PROBLEM 6.1 *Mass Haul Diagram*

The below Mass Haul Diagram belongs to a wide trapezoidal seawater canal project that has a soil shrinkage factor of 14.5%

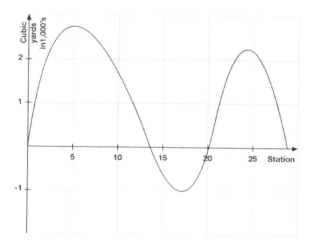

Considering a Free Haul Distance of 500 ft and a Free Haul Volume of 1,600 $yard^3$, wastage in $yard^3$ for this project is most nearly:

(A) 800

(B) 400

(C) 1,400

(D) 685

PROBLEM 6.2 *Volume of Excavation*

The below cut and fill diagram belongs to a highway project. The project's existing ground consists of soil with 20% shrinkage factor and 15% swell factor.

The diagram's negative y-axis represents cross section areas to be cut and the positive y-axis represents cross areas sections to be filled.

The project shall balance its cut and fill material and the rest goes to waste.

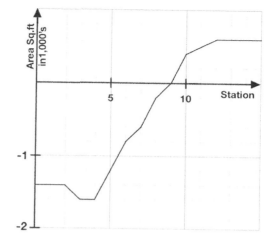

Considering dump trucks can haul up to 11 $yard^3$ per trip, the number of trips expected to haul waste outside the project is most nearly:

(A) 2,470

(B) 2,190

(C) 1,240

(D) 2,680

PROBLEM 6.3 *Ground Level Uphill*

The left-hand-side staff represents a forward sight reading of 3.25 ft and the back sight reading at level +691.85 is 7.85 ft.

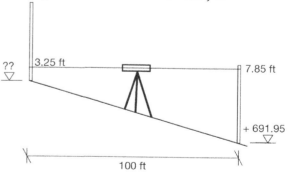

The missing level uphill is most nearly:

(A) 696.55

(B) 699.80

(C) 693.30

(D) 703.05

PROBLEM 6.4 *Soil Loss Prevention*

The following bare land properties are provided for soil loss calculation and prevention:
- 300 ft long sloped at 2%
- Type of soil is Sandy Loam with 0.5% organic matter
- Rainfall intensity index is 200 $ton.in/acre.yr$

Given a permissible soil loss limit of 11 $tons/hectare.yr$, the conservation factor should be:

(A) 0.3

(B) 0.74

(C) 0.014

(D) 0.033

PROBLEM 6.5 *Soil Erodibility*

Which of the following contributes to the amount of soil loss caused by erosion:

I. Porosity: porosity affects soil structure. The more porous the soil, the weaker its structure becomes, leading to increased erodibility.
II. Reduction in vegetation cover contributes to erosion.
III. An increase in kinetic energy from both wind and rainfall leads to higher erosion rates.
IV. Runoff distance plays a role. Longer runoff routes (i.e., longer lands) result in reduced erosion.
V. Runoff slope. The steeper the land slope, the greater the expected erosion.

(A) I + II + III + IV + V

(B) II + III + V

(C) I + II + III + V

(D) II + III + IV + V

PROBLEM 6.6 *Safety Incidence Rate*

The number of safety incidents on a 2 *year* construction project with 150 workers spending 260 *hrs* per month along with an absence rate of 15% is 250.

Half the workers moved into a new 1.5 *year* project with the same working conditions.

Assuming 25% improvement on safety, the expected number of incidents in the second project is most nearly:

(A) 117

(B) 70

(C) 150

(D) 120

PROBLEM 6.7 *Excavation Construction Safety*

A mobile crane is required to lower pipes into the trench shown below. The distance from the center of rotation of the boom to the edge of excavation is 7.5 ft. Also, the center of rotation is 7 ft above ground.

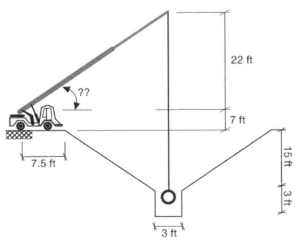

Knowing that the soil type is C per OSHA's definition of soil types, the boom angle is most nearly:

(A) 42.6°

(B) 34.9°

(C) 32.3°

(D) 44.0°

PROBLEM 6.8 Construction Activities

A 12 in thick subgrade layer is being placed for an under-construction 11 ft wide road. Work should be undergoing at a rate of 6.0 miles per day (*).

Consider an average roller/compacter speed of 3.5 mph, 5 passes per roller to achieve required density and a driver efficiency of 50 minutes per hour.

How many 66 in wide rollers are required for this job site to keep up with the stated rate of production:

(A) 1

(B) 2

(C) 3

(D) 4

(*) Consider that this operation runs on a 10 hours shift.

PROBLEM 6.9 Construction Operation Over a Single Footing

During an operation of removing soil from over an embedded foundation for an existing building, the following can happen:

(A) Stress relief as the load which was exerted by the soil on top is removed.

(B) Bearing capacity reduction with possible shear failure.

(C) Reduction in stresses at the bottom reinforcements of the footing.

(D) None of the above.

PROBLEM 6.10 Drainage Channel Cross Section

The below Drainage channel is designed to collect surface water runoff and drain it to its proper outlet. The clear zones for this section have been set out as shown.

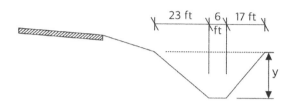

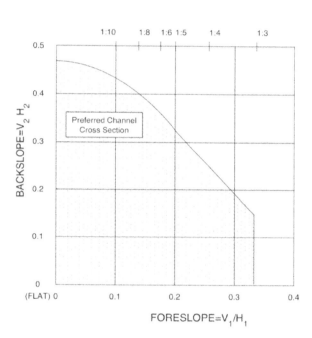

Using the above figure which represents the preferred cross sections for channels with gradual slope changes, the recommended depth for this channel $'y'$ that produces the most desirable drainage cross section is most nearly:

(A) 4.6 ft

(B) 5.8 ft

(C) 6.9 ft

(D) 8.1 ft

PROBLEM 6.11 *Basic Horizontal Curve*

The below is a plan view for an 11 ft wide road with a horizontal curve passing through its centerline and an obstruction as shown. The setback from this obstruction to the edge of the road should not be less than 12 ft.

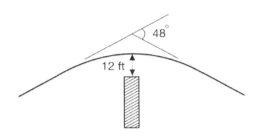

Based on this, the radius, and the length of the curve in ft should nearly be:

(A) 138 & 116

(B) 36 & 31

(C) 53 & 44

(D) 200 & 170

PROBLEM 6.12 *Basic Vertical Curve*

The below vertical curve has a 14 ft bridge passing over the curve's PVI as shown:

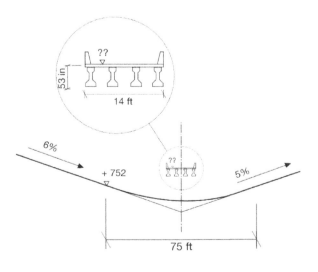

With a 16 ft clearance, and asphalt level at PVC of +752, bridge deck level should be designed at:

(A) +750.9

(B) +771.3

(C) +771.0

(D) +769.9

PROBLEM 6.13 *Points on Vertical Curve*

The following points fall on a vertical curve profile:

Point	Station	Elevation
PVC	0 + 025	72.5
PVI	0 + 275	65
A point on the curve	0 + 150	70

Using the above data, the initial and final grade for this section are as follows:

(A) −0.03, 0.05

(B) −0.03, 0.06

(C) −0.06, 0.10

(D) −0.06, 0.08

PROBLEM 6.14 *Crest Curve Slope*

The below crest curve starts at a PVC of 1 + 00 with an upward going slope of (+4%). The PVI is located at 5 + 00 with a level of +100 after which it heads into a downward slope of (−3%).

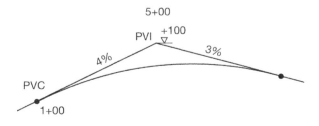

Given the above information, a tangent sloped with (−1.5%) gradient is located at the following station:

(A) 6 + 35

(B) 4 + 14

(C) 7 + 29

(D) 8 + 40

PROBLEM 6.15 *Lowest Point on Sag Curve*

The below sag curve's PVC starts at station 1 + 20 as shown below, elevation at PVI is +77 with a downgrade of 4% and an upgrade of 3%.

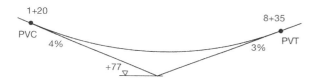

Given the above information, the elevation of the lowest point on this curve is most nearly:

(A) +83

(B) +82

(C) +80

(D) +85

(⁂) PROBLEM 6.16 *Retaining Wall Safety Factor*

The below reinforced concrete cantilever retaining wall has the following properties:

- Concrete density $\gamma_{concrete} = 150\ pcf$
- Soil density $\gamma_{soil} = 130\ pcf$
- Water density $\gamma_{water} = 62.4\ pcf$
- Soil's friction angle $\emptyset'_{soil} = 37°$
- Ground water level 6 ft below surface

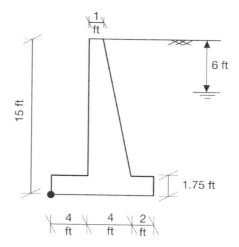

Based on the above information, wall overturning Safety Factor (*) is most nearly:

(A) 2.1

(B) 2.7

(C) 3.7

(D) 15.9

* Consider overturning will occur around left most bottom concrete edge.

PROBLEM 6.17 *Retaining Wall Applicable Loads*

The figure below is for a plain concrete gravity retaining wall with an 8 ft Clayey Silt backfill and base.

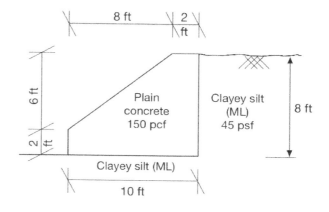

Considering that the cohesion exhibited by the clayey silts to the bottom of the wall is 130 psf, Safety Factors against sliding and overturning for this wall are as follows:

(A) 0.9 for sliding and 13.2 for overturning.

(B) 1.1 for sliding and 1.3 for overturning.

(C) 2.9 for sliding and 13.2 for overturning.

(D) More information required.

PROBLEM 6.18 Construction Methods

Choose three from the below list of construction machinery used in asphalting:

☐ Motor grader

☐ Paver screed

☐ Pneumatic tire roller

☐ Milling Machine

☐ Double steel roller

☐ Single smooth drum roller

PROBLEM 6.19 Balancing a Free Body

The below crane lifts a steel L-shaped object and the object tilts to the left as shown with an angle θ with the vertical line.

The crane's cable is represented by a hinged connection as shown in the sketch below and to the right.

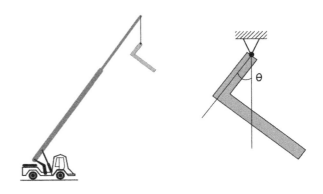

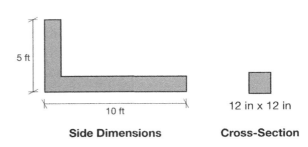

Side Dimensions   Cross-Section

Given the solid steel object dimensions shown above, the angle with the vertical line θ is most nearly:

(A) $35°$

(B) $40°$

(C) $45°$

(D) $47.5°$

📖 PROBLEM 6.20 Subbase Stabilization

A road contractor is requested to stabilize the subbase layer before the construction work starts as it includes moderates amount of clay gravel soils within its particles.

Considering that the *Plasticity Index* (PI) for the affected layer is '40', the best improvement method is as follows:

(A) Cement stabilization by mixing 3% Portland cement with the subbase material.

(B) Cement stabilization by mixing 9% Portland cement with the subbase material.

(C) Apply a small percentage of lime, typically 0.5% to 3%, to the affected material with a process called lime modification.

(D) Lime stabilization by mixing nearly 3% to 5% of lime to the affected layer.

SOLUTION 6.1

The Free Haul Distance (FHD) has been given in the problem as $500\,ft$. The FHD is the distance below which earthmoving is considered part of the contract and contractor cannot claim for extras for overhauling.

To identify stations which fall within the FHD, a $500\,ft$ to-scale horizonal line is drawn and fit in position to intersect close to the peaks and troughs of the Mass Haul Diagram (MHD) curves as shown in the figure below. The y-axis generated values from the FHD intersection with the MHD curves represent the quantity which will be hauled as part of the contract price with no extras.

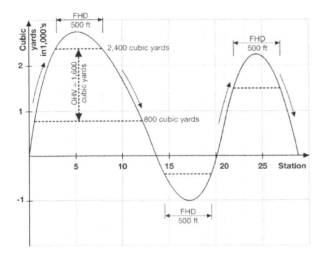

The Over Haul Volume (OVH) is the volume beyond which earthmoving can be claimed as extra by the contractor. OVH is the vertical distance from the FHD towards the x-axis (whether upwards or downwards) to an imaginary horizontal line that intersects with the two sides of the semi parabolic curves.

In this case, and as given in the question, the $OHV = 1,600\,yard^3$ measured as a vertical distance from the FHD down to an intersection of $800\,yard^3$ for the first MHD curve. This does not apply to the second or the third curves because there isn't enough vertical distance to establish such ordinates.

The vertical ordinates generated form this intersection represent either: (1) the waste material when the curve is moving upwards (i.e., cut sections which is more economical to dump outside the project), or (2) the borrow material when the curve is moving downwards (i.e., fill sections which is more economical to supply its material from outside the project). The horizonal intersection is hence called the Limit of Economic Hall (LEH). Shrinkage does not apply in the cut section therefore, wastage in this case will only be $800\,yard^3$ (*).

Correct Answer is (A)

(*) To put things into perspective for this question, and as concluded from the MHD, the material that should go to waste is located between stations $'0+00'$ and $'0+70'$ of the canal that is to be built. If we assume this is a $60\,ft$ wide canal construction project, the height of this cut section that should go to waste is nearly:

$$\frac{800\,yard^3 \times 27\,\frac{ft^3}{yard^3}}{60\,ft \times 70\,ft} \approx 5\,feet$$

SOLUTION 6.2

The average area method is used to measure the volume for each two consecutive cross-sectional areas where the distance between consecutive areas is a station = $100\,ft$:

$$V = L\left(\frac{A_1 + A_2}{2}\right)$$

The following two tables represent cross-sections measured at each station along with the volume between every two consecutive stations.

Bank (undisturbed) cut volume $V_{B,cut} = -950,000\ ft^3$ is converted to lose volume $V_{L,cut}$. Fill requirements $V_{B,fill}$ is also converted into loose volume $V_{L,fill}$. The balance of both goes to waste.

$$V_{L,cut} = \left(1 + \frac{S_w}{100}\right) V_{B,cut}$$

$$= \left(1 + \frac{15}{100}\right)(-950,000)$$

$$= -1,092,500\ ft^3$$

$$V_{L,fill} = \left(1 + \frac{S_h}{100}\right) V_{B,fill}$$

$$= \left(1 + \frac{20}{100}\right)(300,000)$$

$$= 360,000\ ft^3$$

Station	Cut Area ft^2	Fill Area ft^2
0+00	−1,400	
1+00	−1,400	
2+00	−1,400	
3+00	−1,600	
4+00	−1,600	
5+00	−1,200	
6+00	−800	
7+00	−600	
8+00	−200	
9+00	0	0
10+00		400
11+00		500
12+00		600
13+00		600
14+00		600
15+00		600

Station	Cut Volume ft^3	Fill Volume ft^3
0+00 to 1+00	−140,000	
1+00 to 2+00	−140,000	
2+00 to 3+00	−150,000	
3+00 to 4+00	−160,000	
4+00 to 5+00	−140,000	
5+00 to 6+00	−100,000	
6+00 to 7+00	−70,000	
7+00 to 8+00	−40,000	
8+00 to 9+00	−10,000	0
9+00 to 10+00	0	20,000
10+00 to 11+00		45,000
11+00 to 12+00		55.000
12+00 to 13+00		60,000
13+00 to 14+00		60,000
14+00 to 15+00		60,000
Total	**−950,000**	**300,000**

$$V_{L,waste} = -1,092,500 + 360,000$$

$$= -732,500\ ft^3$$

$$No.\ of\ trips = \frac{732,500\ ft^3}{11\ yd^3 \times 27 \frac{ft^3}{yd^3}}$$

$$= 2,466\ trip$$

Correct Answer is (A)

SOLUTION 6.3

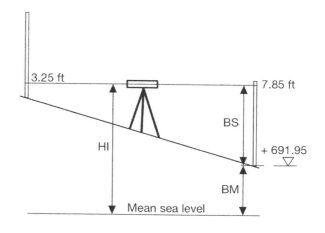

In reference to the above sketch, start with establishing the mean sea level as shown and determine the height of instrument (HI), based upon which the missing level can be determined as follows:

$$HI = BM + BS$$
$$= 691.95 + 7.85$$
$$= 699.8$$
$$Level = HI - FS$$
$$= 699.8 - 3.25$$
$$= 696.55$$

Correct Answer is (A)

SOLUTION 6.4
The *NCEES Handbook version 2.0,* Chapter 6 Water Resources and Environmental, Section 6.5.9.2 Erosion/ Revised Universal Soil Loss equation of page 411 is referred to.

$$A = R.K.LS.C.P$$

P is the conservation factor, given all inputs of this equation are provided in the body of the question, P is calculated as follows:

$$P = \frac{A}{R.K.LS.C}$$

A is the amount of soil loss due to erosion measured in *tons per acre per year*:

$$A = 11 \frac{tons}{hectare.yr} \times \frac{1\ hectare}{2.47\ acre}$$
$$= 4.45 \frac{tons}{acre.yr}$$

K is the soil erodibility factor taken from the same section of *NCEES Handbook version 2.0* page 412 as '0.27' for Sandy Loam with 0.5% organic matter.

LS is the topographic factor taken from the same section of *NCEES Handbook version 2.0* page 412 as '0.28' for a 300 ft land sloped at 2%.

C is the crop and cover management factor taken as '1.0' for bare land.

$$P = \frac{A}{R.K.LS.C}$$
$$= \frac{4.45}{200 \times 0.27 \times 0.28 \times 1.0}$$
$$= 0.29\ (*)$$

Correct Answer is (A)

(*) A conservation value of $P = 0.29$ requires strip cropping and contour farming. In a nutshell this requires growing crops in a systematic arrangement of strips along the contours and across a sloping field.

SOLUTION 6.5
The *NCEES Handbook version 2.0,* Chapter 6, Section 6.5.9.2 Erosion/ Revised Universal Soil Loss equation is referred to in order to provide context into the solution.

The Revised Universal Soil Loss equation:

$$A = R.K.LS.C.P$$

A is the amount of soil loss due to erosion (*tons per acre per year*), R is the rainfall erosion index or the climatic erosivity, K is soil erodibility factor, LS is the topographic factor and is taken from the table provided in page 412 of the handbook's version 2.0. C represents vegetation and is called the crop and cover management factor, and P is the erosion control practices factor.

Factors Influencing Soil Loss:

Rainfall erosion index R considers the intensity, duration, and continuity of rainfall. Rainfall erosivity and its relationship to

kinetic energy play a crucial role in erosion.
This makes Statement III true.

Soil erodibility K is its susceptibility to erosion. Factors affecting K include soil aggregation and structure. The more porous the soil, the reduced runoff it shall experience and the lesser effect it would have on its continuous erodibility.
This marks Statement I as incorrect.

In a similar fashion, vegetation cover acts as a barrier against erosion by obstructing water velocity, hence lesser runoff is experienced with more cover.
This makes Statement II true.

Topographic factor LS, specifically the slope length and its steepness, are both proportional to erosion. Which means, more length and more slope causes more erosion.
This marks Statement IV as incorrect, however, it makes Statement V true.

Correct Answer is (B)

SOLUTION 6.6
Safety Incidence Rate IR equation can be picked up from the *NCEES Handbook* Section 2.6.1.1.

$$IR = \frac{N \times 200,000}{T}$$

$$IR_{1st\ project} = \frac{250 \times 200,000}{0.85 \times 2 \times 12 \times 260 \times 150}$$

$$= 62.85$$

$$N_{2nd\ project} = \frac{IR_{first\ project, improved} \times T}{200,000}$$

$$= \frac{(0.75 \times 62.85) \times 0.85 \times 1.5 \times 12 \times 260 \times 75}{200,000}$$

$$= 70$$

Correct Answer is (B)

SOLUTION 6.7
The *NCEES Handbook,* Chapter 3 Geotechnical, Section 3.10 Trench and Excavation Construction Safety is referred to.

Soil type C should be excavated with a slope of 1 vertical to 1 ½ horizontal as shown below:

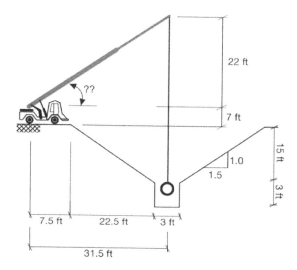

With the above slope configuration, all missing dimensions can be determined, and the required angle can be calculated using trigonometry as follows:

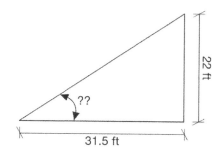

$$\emptyset = arctan\left(\frac{22}{31.5}\right) = 34.9°$$

Correct Answer is (B)

SOLUTION 6.8

The below equation can be found in the *NCEES Handbook* Section 2.3.3.1.

Compacted cubic yards per hour:

$= \frac{1}{n}(16.3 \times W \times S \times L \times efficiency)$

$= \frac{1}{5}\left(16.3 \times \frac{66\,in \times 1\,ft}{12\,in} \times 3.5\,mph \times 12 \times \frac{50\,min}{60\,min}\right)$

$= 628\ yard^3/hour\ (6{,}280\ yard^3/day)$

A 6 *mile*, 12 *in* thick, 11 *ft* wide stretch has the following volume:

$\frac{\left(6\,mile \times 5{,}280\,\frac{ft}{mile}\right) \times \left(12\,in \times \frac{1\,ft}{12\,in}\right) \times 11\,ft}{27\,\frac{ft^3}{yard^3}} = 12{,}903\ yd^3$

$No.\ of\ rollers\ req'd = \frac{12{,}903}{6{,}280} \cong 2$

Correct Answer is (B)

SOLUTION 6.9

The *NCEES Handbook*, Chapter 4 Geotechnical, Section 3.4 Bearing Capacity can be referred in this question.

Bearing capacity equations of Section 3.4.2 for strip footings – copied below for ease of reference – which are fundamentally similar to bearing capacity equations for other footing types, has the total surcharge pressure at the base of the footing q as part of, and a major contributor to, the bearing capacity of the soil beneath. See below:

$q_{ult} = c(N_c) + q(N_q) + 0.5\gamma(B_f)(N_\gamma)$

$q = q_{app} + \gamma_a D_f$

Where q_{app} is the surcharge pressure at surface, which also has a positive impact on bearing capacities. γ_a is density of soil above the base of the footing, and D_f is the depth of the footing.

In conclusion, the removal of soil, or any other surcharge load on top of embedded foundations, by either excavation or scour, can substantially reduce the ultimate bearing capacity and may cause a catastrophic shear failure.

Correct Answer is (B)

SOLUTION 6.10

The given figure is used to determine the preferred cross section for this channel given. Copied below for ease of reference.

The preferred cross section should fall inside the shaded area of the graph. To determine this, Foreslopes and Backslopes for each proposed depth are calculated and plotted as shown above.

The following calculation is performed on option (A) which happens to be the only option that fits this criterion.

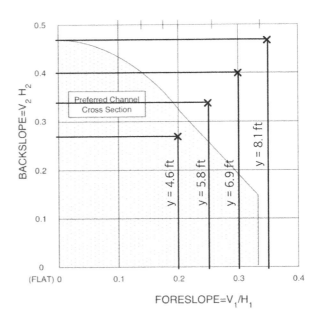

$$Foreslope = \frac{4.6}{23} = 0.2$$

$$Backslope = \frac{4.6}{17} = 0.27$$

Correct Answer is (A)

SOLUTION 6.11

The requested setback is identified in the *NCEES Handbook version 2.0* as distance M and is measured to the center of the road as identified in the question.

$$M = R - R\cos\left(\frac{\Delta}{2}\right)$$

$$(12 + 0.5 \times 11)\,ft = R \times \left(1 - \cos\left(\frac{48}{2}\right)\right)$$

$$\rightarrow R = 202.4\,ft$$

$$L = \frac{R\,\Delta\,\pi}{180} = \frac{202.4 \times 48 \times \pi}{180} = 169.5\,ft$$

Correct Answer is (D)

SOLUTION 6.12

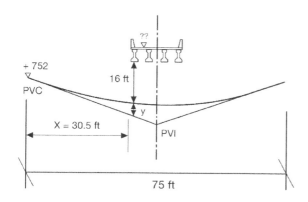

$Y_{PVC} = +752$

$Curve\ elevation = Y_{PVC} + g_1 x + a x^2$

$a = \frac{g_2 - g_1}{2L} = \frac{0.05 - (-0.06)}{2 \times 75} = 7.33 \times 10^{-4}$

$'x'$ is calculated to the edge of the bridge:

$$x = \frac{75}{2} - \frac{14}{2} = 30.5\,ft$$

Curve Elevation

$$= 752 + (-0.06) \times 30.5$$
$$+ 7.33 \times 10^{-4} \times 30.5^2$$
$$= +750.85$$

Bridge Deck Elevation

$$= 750.85 + 16 + \frac{53}{12} = +771.3$$

Correct Answer is (B)

SOLUTION 6.13

In reference to the *NCEES Handbook version 2.0*, Section 5.3.1 Symmetrical Vertical Curve Formula, and the table given in the question – copied here for reference, a vertical curve can be constructed as follows:

Point	Station	Elevation
PVC	0 + 025	72.5
PVI	0 + 275	65
A point on the curve	0 + 150	70

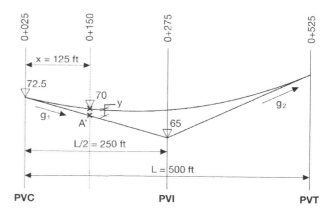

$$g_1 = \frac{65 - 72.5}{275 - 25} = -0.03$$

Determine Elevation of the point on the Back Tangent at station $0 + 150$ located at $x = 125\,ft$.

$A'_{Elevation} = 72.5 - 0.03 \times 125 = 68.75$

$y = 70 - 68.75 = 1.25$

$y = ax^2$

$\rightarrow a = \frac{y}{x^2} = \frac{1.25}{125^2} = 8 \times 10^{-5}$

$a = \frac{g_2 - g_1}{2L}$

$\rightarrow g_2 = 2aL + g_1$

$= 2 \times 8 \times 10^{-5} \times 500 + (-0.03)$

$= 0.05$

Correct Answer is (A)

SOLUTION 6.14
The *NCEES Handbook version 2.0* Chapter 5 Transportation and the vertical curve equations are referred to in this solution.

Start with constructing the curve's parabolic equation as defined in Section 5.3.1 as follows:

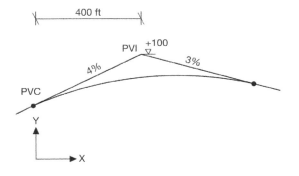

$y = Y_{PVC} + g_1 x + ax^2$

$Y_{PVC} = 100 - 0.04 \times 400 = 84$

$a = \frac{g_2 - g_1}{2L} = \frac{-0.07}{1600} = -4.375 \times 10^{-5}$

Substituting the above back into the parabolic equation:

$y = 84 + 0.04x + (-4.375 \times 10^{-5})x^2$

Gradient of the curved slope is represented by the first derivative of the above equation which is as follows:

$\frac{dy}{dx} = 0.04 + (-8.75 \times 10^{-5})x$

$-0.015 = 0.04 - (8.75 \times 10^{-5})x$

$x = 628.6 \, ft$

$Station_{\frac{dy}{dx} = -1.5\%} = [1 + 00] + (628.6 \, ft)$

$= 7 + 28.6$

Correct Answer is (C)

SOLUTION 6.15
The *NCEES Handbook version 2.0* Chapter 5 Transportation and the vertical curve equations are referred to in this solution.

Per Section 5.3.1, the lowest (or highest point) is located at x_m which is calculated as follows:

$x_m = -\left(\frac{g_1}{2a}\right)$

$a = \frac{g_2 - g_1}{2L} = \frac{0.03 - (-0.04)}{2(835 - 120)} = 4.9 \times 10^{-5}$

$x_m = -\left(\frac{-0.04}{2 \times 4.9 \times 10^{-5}}\right) = 408.6 \, ft$

Construct the equation of the curve to determine the required elevation at $x = x_m = 408.6 \, ft$ as follows:

$y = Y_{PVC} + g_1 x + ax^2$

$Y_{PVC} = 77 + 0.04 \times \frac{835 - 120}{2} = 91.3$

$y = 91.3 - 0.04x + 4.9 \times 10^{-5} x^2$

$y_{@408.6} = 91.3 - 0.04(408.6)$

$\qquad + 4.9 \times 10^{-5}(408.6)^2$

$= +83.14$

Correct Answer is (A)

(✱) SOLUTION 6.16
Based on the information given in the question, Rankine's active coefficient is calculated as follows:

$k_a = tan^2\left(45 - \frac{\phi'}{2}\right)$

$\quad = tan^2\left(45 - \frac{37°}{2}\right) = 0.25$

The resultant soil pressure equation:

$p_a = \frac{k_a h^2 \gamma_{soil}}{2}$ triangular shaped pressure

$p_a = k_a\, h_{above\,water}\, h_{below\,water} \gamma_{soil}$
$\qquad\qquad$ rectangular shaped pressure

Based on this, lateral/overturning pressures are calculated per linear ft as follows:

Pressure resultant force from normal weight of soil:

$p_{a,(0-6ft)} = \frac{1}{2} \times 0.25 \times (6ft)^2 \times 130 \frac{lb}{ft^3}$

$\qquad\quad = 585\ lb/ft$

$p_{a,(6-15ft)} = 0.25 \times 6 \times 9 \times 130 \frac{lb}{ft^3}$

$\qquad\quad = 1{,}755\ lb/ft$

Pressure resultant force from effective weight of soil:

$p_{a,(6-15ft)} = \frac{1}{2} \times 0.25 \times (9\,ft)^2$

$\qquad\qquad \times (130 - 62.4) \frac{lb}{ft^3}$

$\qquad\quad = 684.5\ lb/ft$

Pressure resultant force from hydrostatic pressures:

$p_{water,(6-15ft)} = \frac{1}{2} \times (9ft)^2 \times 62.4 \frac{lb}{ft^3}$

$\qquad\qquad = 2{,}527.2\ lb/ft$

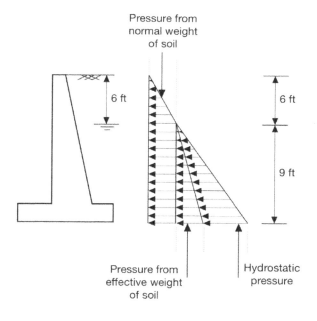

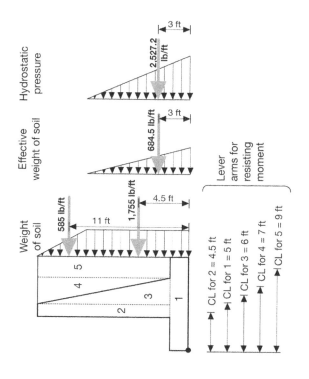

Overturning moments (O.M.) around the marked left most point:

Description	Force lb	Lever arm ft	Overturning Moment lb.ft
Soil 0-6 ft	585.0	11.0	6,435.0
Soil 6-15 ft	1,755	4.5	7,897.5
Eff. soil 6-15 ft	684.5	3.0	2,053.5
Hydrostatic 6-15 ft	2,527.2	3.0	7,581.6
			23,967.5

The sum of overturning moments:
$\sum O.M. = 23,967.5 \ lb.ft/ft$

Resisting moments (R.M.) with items 4 and 5 belong to the soil, $\gamma_{soil} = 130 \ pcf$:

Sec	Area ft^2	Weight lb	Lever arm ft	Resisting Moment lb.ft
1	17.5	2,625.0	5.0	13,125.0
2	13.25	1,987.5	4.5	8,943.8
3	19.9	2,985.0	6.0	17,910.0
4	19.9	2,587.0	7.0	18,109.0
5	26.5	3,445.0	9.0	31,005.0
				89,092.8

The sum of resisting moments
$\sum R.M. = 89,092.8 \ lb.ft/ft$

Safety Factor calculation:
$S.F._{OT} = \frac{\sum R.M.}{\sum O.M.} = \frac{89,092.8}{23,967.5} = 3.7$

Correct Answer is (C)

SOLUTION 6.17
Overturning and sliding forces along with the resisting forces are calculated as follows:

Overturning:
Safety Factor against overturning is determined by calculating the Overturning Moment $\sum O.M.$ and the Resisting Moment $\sum R.M.$ around point 'O' as follows:

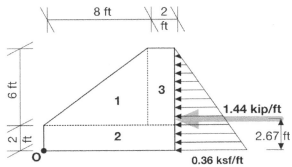

Resisting Moments $\sum R.M.$ as exhibited by the concrete wall:

Section	Volume ft^3/ft	Wt. kip/ft	Lever arm ft	R.M. kip.ft/ft
1	24	3.6	5.33	19.2
2	20	3.0	5.0	15.0
3	12	1.8	9.0	16.2
Totals	56	8.4		50.4

$\sum R.M. = 50.4 \ kip.ft/ft$

$\sum O.M. = [0.5 \times (45 \ psf/ft \times 8 \ ft) \times 8 \ ft] \times 2.67 \ ft$

$= 3,845 \ lb.ft/ft \ (3.8 \ kip.ft/ft)$

$SF_{OT} = \frac{\sum R.M.}{\sum O.M.} = \frac{50.4}{3.8} = 13.26$

Sliding:
Safety factor against sliding is determined by calculating the sliding force F_S and resisting force F_R as follows:

$F_S = 0.5 \times (45 \ psf/ft \times 8ft) \times 8ft \times \frac{1 \ kip}{1,000 \ lb}$

$= 1.44 \ kip/ft$

Resisting force is give in the question as 130 psf and is used/multiplied by the bottom area of the wall:

$F_R = 130 \ psf \times 1 \ ft \times 10 \ ft \times \frac{1 \ kip}{1,000 \ lb}$

$= 1.3 \ kip/ft$

$SF_{sliding} = \frac{F_R}{F_S} = \frac{1.3}{1.44} = 0.9$

Correct Answer is (A)

SOLUTION 6.18

This question requires more of a hands-on field experience and is not a typical question that can be found in a handbook.

Construction machinery used for the purpose of asphalting are as follows:

✓ *(Double) Steel wheel roller*
Steel wheel rollers are self-propelled compaction devices that use steel drums to compress hot asphalt mixes. Usually one, two or three drums. Two drums are most used.

✓ *Paver screed*
Paver screed receives the hot mix from dumping trucks, distributes it and paves it to the required thickness. Pavers have enough power to push the truck as it empties its content into the paver's receiving shaft. Pavers control the speed by which the truck moves, and it can achieve 70-80% density of the layer.

✓ *Pneumatic tire roller*
Those are self-propelled compaction devices that have pneumatic tires which provide proper and smooth compaction to the layer beneath as those tires have no threads in them.

The following machinery are not commonly used in a normal asphalting operation:

✗ *Milling machine*
Milling machines are used to cut existing asphalt or remove a top layer of the existing pavement to create a bed surface. In this case it cannot be claimed to be useful during an asphalting operation unless specifically required on the field.

✗ *Single smooth drum roller*
Smooth drum rollers use static and vibratory pressure to compact rough material such as gravel, rocks, and sand. They are more effective in granular material, and they would not normally produce a smooth surface. A single drum roller has tires at the back and those tires come with deep threads which are not suitable for finishing asphalt layers.

✗ *Motor grader*
Motor graders have an adjustable blade that can be used to complete various construction activities such as surface leveling, fine grading, creating slopes, creating ditches and earth moving. Graders have tires with deep threads which makes it challenging to work on top of a hot mix asphalt. Graders can however be used for asphalting during abnormal circumstances, such as during a breakdown of the asphalt paver, or paving steep slopes, or locations with narrow access that screeds cannot enter.

SOLUTION 6.19

The principle of static equilibrium is used to calculate the final position of a hanging object or a free body.

The force of gravity which passes through the object's Center of Gravity CG is balanced by the tension in the cable. And provided there is only one unrestrained point of contact or support, there will be no moment generated from this arrangement, which means that the cable's extended virtual line should pass through the object's CG.

First, start with calculating the location of center of gravity CG by taking the datum at its bottom left corner as shown below:

$$Wt_1 = \left(10\,ft \times \frac{12\,in}{ft}\right) \times 12\,in \times 12\,in \times \gamma_{steel}$$
$$= 17{,}280\,\gamma_{steel}$$

$$Wt_2 = \left(5\,ft \times \frac{12\,in}{ft} - 12\right) \times 12\,in \times 12\,in \times \gamma_{steel}$$

$$= 6{,}912\,\gamma_{steel}$$

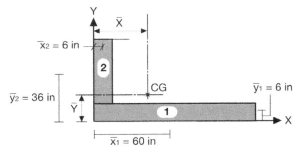

$$\bar{X} = \frac{Wt_1 \bar{x}_1 + Wt_2 \bar{x}_2}{Wt_1 + Wt_2}$$

$$= \frac{17{,}280\,\gamma_{steel} \times 60\,in + 6{,}912\,\gamma_{steel} \times 6\,in}{17{,}280\,\gamma_{steel} + 6{,}912\,\gamma_{steel}}$$

$$= 44.6\,in\;(3.7\,ft)$$

$$\bar{Y} = \frac{Wt_1 \bar{y}_1 + Wt_2 \bar{y}_2}{Wt_1 + Wt_2}$$

$$= \frac{17{,}280\,\gamma_{steel} \times 6\,in + 6{,}912\,\gamma_{steel} \times 36\,in}{17{,}280\,\gamma_{steel} + 6{,}912\,\gamma_{steel}}$$

$$= 14.6\,in\;(1.2\,ft)$$

Second, tilt the object as shown below until the vertical line from the crane's cable vertically aligns with the CG. Construct a right angle triangle and determine the requested angle using trigonometry as follows:

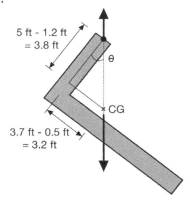

$$\theta = \arctan\left(\frac{3.2\,ft}{3.8\,ft}\right) = 40.1°$$

Correct Answer is (B)

SOLUTION 6.20

Cement stabilization is considered when *plasticity index* is less than 10 and is used to strengthen granular soils by mixing in Portland cement, typically $3 - 5\%$ of the soil dry weight. Check *NCEES Handbook* Section 2.5.4 for more information.

Lime modification is used to improve fine grained soils with the addition of $0.5\%\;to\;3\%$ to the soil dry weight.

However, with a plasticity index > 10 for subbase or base materials with moderate and "predominant" amount of clay gravel soils, lime stabilization is considered the best option. Typically, $3\%\;to\;5\%$ should be enough to dry up the mud contained in the subbase layer.

Correct Answer is (D)

REFERENCES, BIBLIOGRAPHY, PERMISSIONS & INDEX

References

The following references have been used throughout the book and they are all required, and you have to obtain and study them thoroughly prior to the exam:

1. NCEES Handbook, 2023. *PE Civil Reference Handbook Version 2.0.* National Council of Examiners for Engineering and Surveying.

2. *The Recommended Standards for Wastewater Facilities, 2014.* Great Lakes - Upper Mississippi River Board of State and Provincial Public Health and Environmental Managers.

3. The *Recommended Standards for Water Works Facilities, 2018.* Great Lakes - Upper Mississippi River Board of State and Provincial Public Health and Environmental Managers.

Bibliography

Throughout this book, we have drawn upon various references to strengthen the theoretical foundations and concepts presented in this book. Given the diverse nature of this field, it is unlikely that all relevant information can be found in a single source. Consequently, we have gathered information from multiple sources to provide you with a good coverage for the water resources and environmental engineering knowledge areas.

While you are not obligated to seek out these references, they serve as valuable resources for anyone interested in deepening their understanding of the topics presented here. Feel free to explore them if you wish to enhance your knowledge in specific areas.

Textbooks:

1. Howe, K.J., Hand, D.W., Crittenden, J.C, Trussell, R.R & Tchobanoglous, G., 2012. *Principle of Water Treatment.* New Jersey: John Wiley & Sons, Inc.

2. Metcalf & Eddy, Tchobanoglous, G., Stensel, H.D, Tsuchihashi, R. & Burton, F. 2013. *Wastewater Engineering: Treatment and Resource Recovery (5th Edition).* New York: McGraw-Hill.

3. Strum, T. W., 2001. *Open Channel Hydraulics.* New York: McGraw-Hill.

Scholarly Journal Articles:

1. Bossard, G.; Briggs, G.; and Stow, T., 1995. *Optimizing Chlorination & Dechlorination at a Wastewater Treatment Plant.* Public Works. 126 (January): 33-5.

2. Barnes, D.G., Dourson, M., Preuss, P., Bellin, J., Derosa, C., Engler, R., Erdreich, L., Farber, T., Fenner-Crisp, P., Francis, E. and Ghali, G., 1988. *Reference dose (RfD): description and use in health risk assessments.* Regulatory toxicology and pharmacology, 8(4), pp.471-486.

3. Nassaji Matin, G., 2022. *Evaluation of effective parameters of Manning roughness coefficients in HDPE culverts via kernel-based approaches.* Journal of Hydroinformatics, 24(6), pp.1111-1126.

4. Hwang, N., Akan, A. & Houghtalen, R., 2016. *Fundamentals of Hydraulic Engineering Systems (5th Edition).* New Jersey: Prentice-Hall.

5. Kumar, A. and Xagoraraki, I., 2010. *Human health risk assessment of pharmaceuticals in water: An uncertainty analysis for meprobamate, carbamazepine, and phenytoin.* Regulatory toxicology and pharmacology, 57(2-3), pp.146-156.

6. Ohanian, E.V., 1992. *New approaches in setting drinking water standards.* Journal of the American College of Toxicology, 11(3), pp.321-324.

7. McCarty, P.L., 1969. *Biological denitrification of wastewaters by addition of organic materials.* In Proceedings for Industrial Waste Conference, Purdue Univ., 1969.

Reports and Guides:

1. Disinfection: CT and Microbial Log Inactivation Calculations, 2009. Colorado Department of Public Health and Environment Water Quality Control Division.

2. 2018 Edition of the Drinking Water Standard and Health Advisory Table. United States Environmental Protection Agency.

3. 2020 The CT Method: A Reference Guide. Developed by the North Carolina Area Wide Optimization Program Team.

4. 2020 Denitrification Profiling and Benchmarking – Technical Guidance Manual. EPA 815-R-20-003. USEPA.

5. 2022 Wastewater Technology Fact Sheer Dechlorination. United States Environmental Protection Agency.

6. USEPA Guidance Manuals for the Surface Water Treatment Rules.

7. U.S. Environmental Protection Agency. LT1ESWTR Disinfection Profiling and Benchmarking: Technical Guidance Manual. Washington, DC: EPA, 2003.

8. Natural Resources Conservation Service. National Engineering Handbook: Part 630 Hydrology. 210-VI-NEH. Washington, DC: U.S. Department of Agriculture, May 2010, Fig. 15-4, p. 15-8.

9. Federal Highway Administration. Hydraulic Design of Highway Culverts: Hydraulic Design Series Number 5. 3rd ed. FHWA-HIF-12-026. Washington, DC: U. Department of Transportation, April 2012.

10. Standard Specifications for Highway Bridges, 2002, by the American Association of State Highway and Transportation Officials, Washington.

11. Roadside Design Guide, 4th edition, 2011 (including February 2012 and July 2015 errata), by the American Association of State Highway and Transportation Officials, Washington.

12. American Society of Civil Engineers. Design and Construction of Sanitary and Storm Sewers. Reston, VA: ASCE, 1970.

Permissions

Figure	Page No. here	Permission
Vertical Stress Contours (Isobars) Based on Boussinesq's Theory **Source**: The Standard Specifications for Highway Bridges, 2002, by the AASHTO	pg. 39 used in Solution 2.1	Republished with Permission from AASHTO
Headwater Depth For Concrete Pipe Culverts with Inlet Control **Source**: Federal Highway Administration. Hydraulic Design of Highway Culverts: Hydraulic Design Series Number 5. 3rd ed.	pg.71 used in Solution 3.20	Republished with Permission from FHWA
Head for Concrete Culverts Flowing Full **Source**: Federal Highway Administration. Hydraulic Design of Highway Culverts: Hydraulic Design Series Number 5. 3rd ed.	pg.72 used in Solution 3.21	Republished with Permission from FHWA
Hydraulic Elements Graph for Circular Pipe (Partial Flow Diagram) **Source**: American Society of Civil Engineers. Design and Construction of Sanitary and Storm Sewers	pg.74 used in Solution 3.25	Publicly Available
Velocity Versus Slope for Shallow Concentrated Flow **Source**: Natural Resources Conservation Service. National Engineering Handbook: Part 630 Hydrology.	pg.86 used in Solution 4.4	Publicly Available
Preferred Cross Sections for Channels with Gradual Slope Changes **Source**: Roadside Design Guide, 4th edition, 2011 (including February 2012 and July 2015 errata)	pg.149 & 158 used in Solution 6.10	Republished with Permission from AASHTO

Index

AASHTO Classification System, 37, 45
Activated Sludge, 109, 110, 112, 119, 132, 133
Activated Sludge Process, 137
Air Stripping, 106, 124
Algae, 141
Anaerobic Sludge Digester, 111
Aquifer Porosity, 88
Aquifer Volume, 88
Bardenpho Process, 138
Basin Design, 84
Bearing Capacity, 35, 43, 158
Benefit Cost Analysis, 18
Benefit/Cost Ratio, 23
Biological Denitrification, 110, 112
Biological Phosphorus Removal, 112, 138
Boussinesq's Method, 39
Carbonate Hardness, 124
Carman-Kozeny Equation, 128
Cement Stabilization, 164
Chezy Equation, 73
Chlorination, 140
Churchill et al. Method, 113, 142
Coefficient of Curvature, 46
Coefficient of Discharge, 120
Coefficient of Uniformity, 46
Composite Curve Number, 84
Concentration Gradient, 127
Confined Aquifer, 87
Conservation of Mass, 115, 116
Consolidation Settlement, 35
Critical Path Method, 22
Critical Slope, 57
Backward Pass, 23
Curb and Gutter Equation, 70
Darcy's Equation, 97
Darcy's Law, 96, 97
Darcy-Weisbach, 49, 61, 65, 111
Dechlorination, 112
Degree of Saturation, 44
Denitrification, 133, 168
Depreciation Rate, 30
Dissolved Air Flotation (DAF), 137
(Double) Steel Wheel Roller, 163
Drag Coefficient, 122
Drinking Water Equivalent Level (DWEL), 114, 144
Dry Density, 43
Dupuit's Equation, 96
Earned Value Management, 22
Forward Pass, 23
Effective Stress, 33, 34, 36, 40, 41, 43, 45
Energy Equation, 60, 61, 79
Energy Loss in Horizontal Hydraulic Jump, 73
Entrance Loss Coefficient, 71
Erosion/ Revised Universal Soil Loss Equation, 156
Exploration Points, 45
Filamentous Bacteria, 110, 131
Flexible Wall Permeameter Test, 38, 46
Flocculator Design, 105
Fluid Power Equations, 67
Food to Microorganism Ratio, 112
Foundation Settlement, 33
Free Haul Distance (FHD), 154
Freundlich Isotherm, 123
Friction Angle, 35, 44
Froude Number, 72, 73
Gantt Chart, 26, 27
General Probability, 94
Giardia Cysts, 108
Gradually Varied Flow, 58, 76
Granular Activated Carbon (GAC), 123
Hagen-Poiseuille Equation, 64
Hardy-Cross Method, 68
Hazen-Williams, 49, 53, 60, 64
Head for Concrete Pipe Culverts Flowing Full, 71
Head Losses, 69, 127
Headwater Depth for Concrete Pipe Culverts, 71
Hydraulic Jump, 56
Hydrograph Development, 85
Constructing a Hydrograph, 94
Hydrologic Budget, 87
Hypochlorous Acid HOCl, 128
IDF Curve Creation, 94
Impulse Momentum Principle, 62
Infiltration/ Horton Model, 95
Inlet Control, 56

Inline Equalization, 117
Isohyetal Method, 83, 89
Jet, 52
Joukowsky's Equation, 65
Limit of Economic Hall (LEH), 154
Liquid Limit, 42, 45
Log Inactivation, 130
Loop Network Flow, 55
Manning Coefficient, 53, 56, 57, 58, 74, 78
Manning Equation, 64, 69, 73, 74, 75
Manning Roughness Coefficient, 70, 71, 74, 78
Mass Haul Diagram (MHD), 147, 154
Membrane Bioreactor (MBR), 132
Membrane Filtration, 126
Mild Slope, 76, 77
Milling Machine, 163
Monod Kinetics, 142
Monod's Model, 143
Moody, Darcy, or Stanton Friction Factor Diagram, 61, 65, 66
Motor Grader, 163
Multiple Aquifer Layers, 98
Natural Resources Conservation Services Method, 92
Net Positive Pressure Suction Head, 67
Net Present Value, 18, 24
Noncarbonate Hardness, 106, 124
Normal Stress, 44
Normally Consolidated Soils, 42
NRCS (SCS) Rainfall Runoff Method, 91
Outlet Control, 56
Overturning Safety Factor, 162
Packed Bed, 53, 66
Parallel Pipe Network, 55, 69
Parshall Flume, 103
Paver Screed, 163
Pipe Flow Depth, 57
Plastic Limit, 42, 45
Plasticity Index, 42, 45, 164
Pneumatic Tire Roller, 163
Porosity, 35, 44, 88, 98, 148
Powdered Activated Carbon (PAC), 106, 123
Precipitation Methods, 83
Present Net Worth, 20, 29

Principle of Static Equilibrium, 163
Pump (Brake) Power, 67
Rainfall Intensity, 86
Rainfall Type II, 92
Rankine Coefficient, 44
Rapid Mixing, 105
Reference Dose RfD, 143
Resources Histogram, 19
Reynolds Number, 56, 61, 65, 70, 127, 128
Rock Quality, 36
Rock Quality Designation, 44
Rose Equation, 128
Rotating Biological Contactor (RBC), 132
Scaling and Affinity Laws, 67
Scour Velocity, 135
Seepage Velocity, 98
Shallow Flow, 83
Shear Strength, 44
Shrinkage Limit, 42, 45
Single Smooth Drum Roller, 163
Sliding Safety Factor, 162
Slope Stability, 34
Snyder Synthetic Unit Hydrograph, 93
Soil Classification System, 37
Soil Erodibility, 148
Specific Retention, 88, 98
Specific Yield, 98
Stream Degradation, 113
Submerged Outlet, 74
Submerged Surfaces and Center of Pressure, 63
Surface Water System Hydrologic Budget, 95, 96
Tailwater Depth, 56, 72
Thiem Equation, 97
Thiessen Method, 83, 89
TMDL, 99, 114, 144
Total Cohesion, 44
Trickling Filter, 132
Unconfined Aquifer, 87
USCS Classification System, 46
Volume of Methane, 135

This page is intentionally left blank

Your Feedback Matters – Make Sure You Share it With Others

Good day,

As you reach the final pages of this book, we would like to express our sincere gratitude for choosing it as your guide to aid you in your journey toward success in the PE exam. We have poured countless hours into meticulously crafting the questions and practice exams within these pages.

Your opinion matters greatly in helping others discover the value of this resource. If you found this book beneficial, kindly consider leaving your positive and honest feedback on the platform that you bought it from - like Amazon. Your words will not only acknowledge the hard work invested into producing this book but will also guide future readers in their quest for quality study materials.

Remember, your review is more than just feedback; it's a beacon for those seeking reliable resources. Your support can make a significant difference, ensuring that this book continues to assist aspiring professionals on their path to success.

Thank you for being a part of this journey, and we appreciate your commitment to sharing your experience with others.

PE ESSENTIAL GUIDES

Made in the USA
Middletown, DE
28 June 2025

77617071R00106